REDSHANK ON WILLOW

HAMPSHIRE DAYS

BY

W. H. HUDSON

AUTHOR OF

"BIRDS AND MAN," "NATURE IN DOWNLAND," ETC.

WITH ILLUSTRATIONS

LONGMANS, GREEN, AND CO
39 PATERNOSTER ROW, LONDON
NEW YORK AND BOMBAY
1903

The greater part of the matter contained in this volume has not appeared before. In the first half of the book use has been made of an article on " Summer in the Forest" from Longman's Maga- *zine ; in the second half I have drawn on articles from the same periodical, on " Wolmer Forest," " A Summer's End on the Itchen," and " Selborne Revisited" ; and I have also made use of an article entitled " A More or Less Happy Family" from the* Badminton Magazine.

CONTENTS

CHAPTER I

CHAPTER II

CHAPTER III

CHAPTER IV

CHAPTER V

CHAPTER VI

CHAPTER X

CHAPTER XI

CHAPTER XII

CHAPTER XIII

CHAPTER XIV

LIST OF ILLUSTRATIONS

xv

J. S., J. Smit; E. H., Mrs. E. Hubbard; A. H. J., A. Heywood
Jones; B. P., Miss Bertha Patmore; D. S, Miss Dora Sulman;
B. H., Bryan Hook; M. G., Miss Marion Gardiner.

Drawings No. 33, 35, and 40 are from photographs by W. T. Green
of Winchester.

HAMPSHIRE DAYS

CHAPTER I

Autumn in the New Forest—Red colour in mammals—November mildness—A house by the Boldre—An ideal spot for small birds—Abundance of nests—Small mammals and the weasel's part—Voles and mice—Hornet and bank-vole—Young shrews —A squirrel's visit — Green woodpecker's drumming-tree— Drumming of other species—Beauty of great spotted wood-pecker—The cuckoo controversy—A cuckoo in a robin's nest— Behaviour of the cuckoo—Extreme irritability—Manner of ejecting eggs and birds from the nest—Loss of irritability— Insensibility of the parent robins—Discourse on mistaken kindness, pain and death in nature, the annual destruction of bird life, and the young cuckoo's instinct.

HERE, by chance, in the early days of December 1902, at the very spot where my book begins, I am about to bring it to an end.

A few days ago, coming hither from the higher country at Silchester, where the trees were already nearly bare, I was surprised to find the oak woods of

A

this lower southern part of the New Forest still in their full autumnal foliage. Even now, so late in the year, after many successive days and nights of rain and wind, they are in leaf still: everywhere the woods are yellow, here where the oak predominates; the stronger golden red and russet tints of the beech are vanished. We have rain and wind on most days, or rather mist and rain by day and wind with storms of rain by night; days, too, or parts of days, when it is very dark and still, and when there is a universal greyness in earth and sky. At such times, seen against the distant slaty darkness or in the blue-grey misty atmosphere, the yellow woods look almost more beautiful than in fine weather.

The wet woodland roads and paths are everywhere strewn, and in places buried deep in fallen leaves—yellow, red, and russet; and this colour is continued under the trees all through the woods, where the dead bracken has now taken that deep tint which it will keep so long as there is rain or mist to wet it for the next four or five months. Dead bracken with dead leaves on a reddish soil; and where the woods are fir, the ground is carpeted with lately-fallen needles of a chestnut red, which brightens almost to orange in the rain. Now, at this season, in this universal redness of the earth where trees and bracken grow, we see that Nature is justified in having given that colour—red and reddish-yellow—to all or to most of her woodland mammals. Fox and foumart and weasel

and stoat; the hare too; the bright squirrel; the
dormouse and harvest-mouse; the bank-vole and the
wood-mouse. Even the common shrew and lesser
shrew, though they rarely come out by day, have a
reddish tinge on their fur. Water-shrew and water-
vole inhabit the banks of streams, and are safer
without such a colour; the dark grey badger is strictly
a night rover.

Sometimes about noon the clouds grow thin in
that part of the sky, low down, where the sun is, and
a pale gleam of sunlight filters through; even a patch
of lucid blue sky sometimes becomes visible for a while:
but the light soon fades; after mid-day the dimness
increases, and before long one begins to think that
evening has come. Withal it is singularly mild. One
could almost imagine in this season of mist and wet
and soft airs in late November that this is a land
where days grew short and dark indeed, but where
winter comes not, and the sensation of cold is unknown. It is pleasant to be out of doors in such
weather, to stand in the coloured woods listening to
that autumn sound of tits and other little birds wandering through the high trees in straggling parties,
talking and calling to one another in their small
sharp voices. Or to walk by the Boldre, or, as some
call it, the Lymington, a slow, tame stream in summer,
invisible till you are close to it; but now, in flood,
the trees that grow on its banks and hid it in summer
are seen standing deep in a broad, rushing, noisy river.

The woodpecker's laugh has the same careless happy sound as in summer: it is scarcely light in the morning before the small wren pours out his sharp bright lyric outside my window; it is time, he tells me, to light my candle and get up. The starlings are about the house all day long, vocal even in the rain, carrying on their perpetual starling conversation —talk and song and recitative; a sort of bird-Yiddish, with fluty fragments of melody stolen from the black-bird, and whistle and click and the music of the triangle thrown in to give variety. So mild is it that in the blackness of night I sometimes wander into the forest paths and by furzy heaths and hedges to listen for the delicate shrill music of our late chirper in the thickets, our Thamnotrizon, about which I shall write later; and look, too, for a late glow-worm shining in some wet green place. Late in October I found one in daylight, creeping about in the grass on Selborne Hill; and some few, left unmarried, may shine much later. And as to the shade-loving grasshopper or leaf cricket, he sings, we know, on mild evenings in November. But I saw no green lamp in the herbage, and I heard only that nightly music of the tawny owl, fluting and hallooing far and near, bird answering bird in the oak woods all along the swollen stream from Brockenhurst to Boldre.

This race of wood owls perhaps have exceptionally strong voices: Wise, in his book on the New Forest, says that their hooting can be heard on a still autumn

evening a distance of two miles. I have no doubt
they can be heard a good mile.

But it is of this, to a bird lover, delectable spot in
the best bird-months of April, May, and June that I
have to write. The house, too, that gave me shelter
must be spoken of; for never have I known any
human habitation in a land where people are discovered dwelling in so many secret, green, out-of-the-
world places, which had so much of nature in and
about it. Grown-up and young people were in it,
and children too, but they were girls, and had always
quite spontaneously practised what I had preached—
pet nothing and persecute nothing. There was no
boy to disturb the wild creatures with his hunting
instincts and loud noises; no dog, no cat, nor any
domestic creature except the placid cows and fowls
which supplied the household with milk and eggs.
A small old picturesque red-brick house with high-
pitched roof and tall chimneys, a great part of it
overrun with ivy and creepers, the walls and tiled
roof stained by time and many-coloured lichen to a
richly variegated greyish red. The date of the house,
cut in a stone tablet in one of the rooms, was 1692.
In front there was no lawn, but a walled plot of
ground with old once ornamental trees and bushes
symmetrically placed — yews, both spreading and
cypress-shaped Irish yew, and tall tapering juniper,
and arbor vitæ; it was a sort of formal garden which
had long thrown off its formality. In a corner of

the ground by the side of these dark plants were laurel, syringa, and lilac bushes, and among these such wildings as thorn, elder, and bramble had grown up, flourishing greatly, and making of that flowery spot a tangled thicket. At the side of the house there was another plot of ground, grass-grown, which had once been the orchard, and still had a few ancient apple and pear trees, nearly past bearing, with good nesting-holes for the tits and starlings in their decayed mossy trunks. There were also a few old ivied shade-trees—chestnuts, fir, and evergreen oak.

Best of all (for the birds) were the small old half-ruined outhouses which had remained from the distant days when the place, originally a manor, had been turned into a farm-house. They were here and there, scattered about, outside the enclosure, ivy-grown, each looking as old and weather-stained and in harmony with its surroundings as the house itself—the small tumble-down barns, the cow-sheds, the pig-house, the granary with open door and the wooden staircase falling to pieces. All was surrounded by old oak woods, and the river was close by. It was an ideal spot for small birds. I have never in England seen so many breeding close together. The commoner species were extraordinarily abundant. Chaffinch and green-finch; blackbird, throstle, and missel-thrush; swallow and martin, and common and lesser whitethroat; garden warbler and blackcap; robin, dunnock, wren, flycatcher, pied wagtail, starling, and sparrow;—one could go round

and put one's hand into half-a-dozen nests of almost any
of these species. And very many of them had become
partial to the old buildings: even in closed rooms
where it was nearly dark, not only wrens, robins, tits,
and wagtails, but blackbirds and throstles and chaffinches were breeding, building on beams and in or
on the old nests of swallows and martins. The hawfinch and bullfinch were also there, the last rearing
its brood within eight yards of the front door. One
of his two nearest neighbours was a gold-crested
wren. When the minute bird was sitting on her
eggs, in her little cradle-nest suspended to a spray of
the yew, every day I would pull the branch down so
that we might all enjoy the sight of the little fairy
bird in her fairy nest which she refused to quit. The
other next-door neighbour of the bullfinch was the
long-tailed tit, which built its beautiful little nest on
a terminal spray of another yew, ten or twelve yards
from the door; and this small creature would also
let us pull the branch down and peep into her well-
feathered interior.

It seemed that from long immunity from persecution,
all these small birds had quite lost their fear of human
beings; but in late May and in June, when many
young birds were out of the nest, one had to walk
warily in the grass for fear of putting a foot on some
little speckled creature patiently waiting to be visited
and fed by its parents.

Nor were there birds only. Little beasties were also

quite abundant; but they were of species that did no
harm (at all events there), and the weasel would come
from time to time to thin them down. Money is paid
to mole-catcher and rat-catcher ; the weasel charges you
nothing: he takes it out in kind. And even as the
jungle tiger, burning bright, and the roaring lion
strike with panic the wild cattle and antelopes and
herds of swine, so does this miniature carnivore,
this fairy tiger of English homesteads and hedges,
fill with trepidation the small deer he hunts and
slays with his needle teeth — Nature's scourge sent
out among her too prolific small rodents; her little
blood-letter who relieves her and restores the balance.
And therefore he, too, with his flat serpent head and
fiery killing soul, is a " dear " creature, being, like the
poet's web-footed beasts of an earlier epoch, " part of
a general plan."

The most abundant of the small furred creatures
were the two short-tailed voles — field-vole and
bank-vole; the last, in his bright chestnut-red, the
prettiest. Whenever I sat down for a few minutes
in the porch I would see one or more run across the
stones from one side, where masses of periwinkle grew
against the house, to the other side, where Virginia
creeper, rose, and an old magnolia tree covered the
wall. One day at the back of the house by the
scullery door I noticed a swaying movement in a
tall seeded stem of dock, and looking down spied a
wee harvest-mouse running and climbing nimbly on

the slender branchlets, feeding daintily on the seed, and looking like a miniature squirrel on a miniature bush.

Just there, close to the door, was a wood-pile, and the hornets had made their nest in it. The year before they had made it in a loft in the house, and before that in the old barn. The splendid insects were coming and going all day, interfering with nobody and nobody interfering with them; and when I put a plate of honey for them on the logs close to their entrance they took no notice of it; but by-and-by bank-voles and wood-mice came stealing out from among the logs and fed on it until it was all gone.

I was surprised, and could only suppose that the hornets did not notice or discover the honey, because no such good thing was looked for so close to their door. Away from home the hornet was quick to discover anything sweet to the taste, and very ready to resent the presence of any other creature at the table.

At the riverside, a few hundred yards from the house, I was sitting in the shade of a large elm tree one day when I was visited by a big hornet, who swept noisily down and settled on the trunk, four or five feet above the ground. A quantity of sap had oozed out into a deep cleft of the rough bark and had congealed there, and the hornet had discovered it. Before he had been long feeding on it I saw a little bank-vole come out from the roots of the tree and run up the trunk, looking very pretty in his bright chestnut fur as he came into the sunlight. Stealing up to the lower end of the cleft full of thickened sap he too began feeding on it. The hornet, who was at the upper end of the cleft, quite four inches apart from the vole, at once stopped eating and regarded the intruder for some time, then advanced towards him in a threatening attitude. The vole was frightened at this, starting and erecting his hair, and once or twice he tried to recover his courage and resume his feeding, but the hornet still keeping up his hostile movements, he eventually slipped quietly down and hid himself at the roots. When the hornet departed he came out again and went to the sap.

Wishing to see more, I spent most of that day and the day following at the spot, and saw hornet and vole meet many times. If the vole was at the sap when the hornet came he was at once driven off, and when the hornet was there first the vole was never allowed to feed, although on every occasion he tried to do so, stealing to his lower place in the gentlest way in order not to give offence, and after beginning to feed affecting not to see that the other had left off eating, and with raised head was regarding him with jealous eyes.

Rarely have I looked on a prettier little comedy in wild life.

But to return to the house. There was quite a happy family at that spot by the back door where the hornets were. A numerous family of shrews were reared, and the young, when they began ex-

BANK-VOLE AND HORNET

ploring the world, used to creep over the white stone by the threshold. The girls would pick them up to feel their soft mole-like fur: the young shrew is a gentle creature and does not attempt to bite. Some of the more adventurous ones were always blundering into the empty flower-pots heaped against the wall, and there they would remain imprisoned until some person found and took them out.

One morning, at half-past four o'clock, when I was lying awake listening to the blackbird, a lively squirrel came dancing into the open window of my bedroom on the first floor. There were writing materials, flowers in glasses, and other objects on the ledge and dressing-table there, and he frisked about among them, chattering, wildly excited at seeing so many curious and pretty things, but he upset nothing; and by-and-by he danced out again into the ivy covering the wall on that side, throwing the colony of breeding sparrows into a great state of consternation.

The river was quite near the house—not half a minute from the front door, though hidden from sight by the trees on its banks. Here, at the nearest point, there was an old half-dead dwarf oak growing by the water and extending one horizontal branch a distance of twenty feet over the stream. This was the favourite drumming-tree of a green woodpecker, and at intervals through the day he would visit it and drum half-a-dozen times or so. This drumming sounded so loud that, following the valley down, I measured the dis-

tance it could be heard and found it just one-third
of a mile. At that distance I could hear it distinctly;
farther on, not at all. It seemed almost incredible
that the sound produced by so small a stick as a
woodpecker's beak striking a tree should be audible
at that distance.

It is hardly to be doubted that the drumming is
used as a love-call, though it is often heard in late
summer. It is, however, in early spring and in the
breeding season that it is oftenest heard, and I have
found that a good imitation of it will sometimes greatly
excite the bird. The same bird may be heard drumming here, there, and everywhere in a wood or copse,
the sound varying somewhat in character and strength
according to the wood; but each bird as a rule has
a favourite drumming-tree, and it probably angers him
to hear another bird at the spot. On one occasion,
finding that a very large, old, and apparently dying
cedar in a wood was constantly used by a woodpecker,
I went to the spot and imitated the sound. Very
soon the bird came and began drumming against me,
close by. I responded, and again he drummed; and
becoming more and more excited he flew close to me,
and passing from tree to tree drummed at every spot
he lighted on.

The other species have the same habit of drumming
on one tree. I have noticed it in the small spotted,
or banded, woodpecker; and have observed that invariably after he had drummed two or three times the

female has come flying to him from some other part of the wood, and the two birds have then both together uttered their loud chirping notes and flown away.

On revisiting the spot a year after I had heard the green woodpecker drumming every day in the oak by the river, I found that he had forsaken it, and that close by, on the other side of the stream, a great spotted woodpecker had selected as his drumming-tree a very big elm growing on the bank. He drummed on a large dead branch about forty feet from the ground, and the sound he made was quite as loud as that of the green bird. It may be that the two big woodpeckers, who play equally well on the same instrument, are intolerant of one another's presence, and that in this case the spotted bird had driven the larger yaffle from his territory.

One of the prettiest spots by the water was that very one where the spotted bird was accustomed to come, and I often went there at noon and sat for an hour on the grassy bank in the shade of the drumming-tree. The river was but thirty to forty feet wide at that spot, with masses of water forget-me-not growing on the opposite bank, clearly reflected in the sherry-coloured sunlit current below. The trees were mostly oaks, in the young vivid green of early June foliage. And one day when the sky, seen through that fresh foliage, was without a stain of vapour in its pure azure, when the wood was full of clear sunlight —so clear that silken spider webs, thirty or forty feet high in the oaks, were visible as shining red and blue and purple lines—the bird, after drumming high above

GREAT SPOTTED WOODPECKER

my head, flew to an oak tree just before me, and cling-ing vertically to the bark on the high part of the trunk, remained there motionless for some time. His statuesque attitude, as he sat with his head thrown well back, the light glinting on his hard polished feathers, black and white and crimson, the setting in which he appeared of greenest translucent leaves and hoary bark and open sunlit space, all together made him seem not only our handsomest woodpecker, but our most beautiful bird. I had seen him at his best, and sitting there motionless amid the wind-fluttered leaves, he was like a bird-figure carved from some beautiful vari-coloured stone.

The most interesting events in animal life observed at this spot relate to the cuckoo in the spring of 1900. Some time before this Dr. Alfred Russel Wallace said, in the course of a talk we had, that he very much wanted me to find out exactly what happened in a nest in which a young cuckoo was hatched. It was, I replied, an old, old story—what could I see, supposing I was lucky enough to find a nest where I could observe it properly, more than Jenner, Hancock, Mrs. Hugh Blackburn, and perhaps other writers, had told us? Yes, it was an old story, he said, and he wanted it told again by some one else. People had lately been discrediting Jenner's account, and as to the other chief authority I had named, one writer, a Dr. Creighton, had said, "As for artists like Mrs. Blackburn, they can draw what they please—all out of their own brains: we can't trust them, or such as them." Sober-minded naturalists had come to regard the habit and abnormal strength attributed to the newly-hatched cuckoo as "not proven" or quite incredible; thus Seebohm had said, "One feels inclined to class these narratives with the equally well-authenticated stories of ghosts and other apparitions which abound."

Since my conversation with Dr. Wallace we have had more of these strange narratives—the fables and ghost stories which the unbelievers are compelled in the end to accept—and all that Dr. Jenner or his assistant saw others have seen, and some observers have even taken snapshots of the young cuckoo in the act of ejecting his fellow-nestling. But it appears from all the accounts which I have so far read, that in every case the observer was impatient and interfered in the business by touching and irritating the young cuckoo, by putting eggs and other objects on his back, and making other experiments. In the instance I am about to give there was no interference by me or by the others who at intervals watched with me.

A robin's nest with three robin's eggs and one of the cuckoo was found in a low bank at the side of the small orchard on May 19, 1900. The bird was in-cubating, and on the afternoon of May 27 the cuckoo hatched out. Unfortunately I did not know how long incubation had been going on before the 19th, but from the fact that the cuckoo was first out, it seems probable that the parasite has this further advantage of coming first from the shell. Long ago I found that this was so in the case of the parasitical Troupials of the genus Molothrus in South America.

I kept a close watch on the nest for the rest of that afternoon and the whole of the following day (the 28th), during which the young cuckoo was lying in the bottom of the nest, helpless as a piece of jelly with a little life in it, and with just strength enough in his neck to lift his head and open his mouth; and then, after a second or two, the wavering head would drop again. At eight o'clock next morning (29th), I found that one robin had come out of the shell, and one egg had been ejected and was lying a few inches below the nest on the sloping bank. Yet the young cuckoo still appeared a weak, helpless, jelly-like creature, as on the previous day. But he had increased greatly in size. I believe that in forty-eight hours from the time of hatching he had quite doubled his bulk, and had grown darker, his naked skin being of a bluish-black colour. The robin, thirty or more hours younger, was little more than half his size, and had a pale, pinkish-yellow skin, thinly clothed with a long black down. The cuckoo occupied the middle of the deep, cup-shaped nest, and his broad back, hollow in the middle, formed a sort of false bottom; but there was a small space between the bird's sides and the nest, and in this space or interstice the one unhatched egg that still remained and the young robin were lying.

During this day (29th) I observed that the pressure

B

of the egg and young robin against his sides irritated
the cuckoo: he was continually moving, jerking and
wriggling his lumpish body this way and that, as if
to get away from the contact. At intervals this irritation would reach its culminating point, and a
series of mechanical movements would begin, all
working blindly but as surely towards the end as if
some devilish intelligence animated the seemingly helpless infant parasite. ·

Of the two objects in the nest the unhatched egg
irritated him the most. The young robin was soft, it
yielded when pressed, and could be made somehow
to fit into the interstice; but the hard, round shell,
pressing against him like a pebble, was torture to
him, and at intervals became unendurable. Then
would come that magical change in him, when he
seemed all at once to become possessed of a preter-
natural power and intelligence, and then the blind
struggle down in the nest would begin. And after
each struggle—each round it might be called—the
cuckoo would fall back again and lie in a state of
collapse, as if the mysterious virtue had gone out of
him. But in a very short time the pressure on his
side would again begin to annoy him, then to tor-
ment him, and at last he would be wrought up to
a fresh effort. Thus in a space of eight minutes I
saw him struggle four separate times, with a period
of collapse after each, to get rid of the robin's egg;
and each struggle involved a long series of move-

ments on his part. On each of these occasions the egg
was pushed or carried up to the wrong or upper side
of the nest, with the result that when the bird jerked
the egg from him it rolled back into the bottom of
the nest. The statement is therefore erroneous that
the cuckoo knows at which side to throw the egg
out. Of course he *knows* nothing, and, as a fact, he
tries to throw the egg up as often as down the
slope.

The process in each case was as follows:—The pressure of the egg against the cuckoo's side, as I have
said, was a constant irritation; but the irritability
varied in degree in different parts of the body. On
the under parts it scarcely existed; its seat was
chiefly on the upper surface, beginning at the sides
and increasing towards the centre, and was greatest
in the hollow of the back. When, in moving, the
egg got pushed up to the upper edge of his side, he
would begin to fidget more and more, and this would
cause it to move round, and so to increase the irritation by touching and pressing against other parts.
When all the bird's efforts to get away from the
object had only made matters worse, he would
cease wriggling and squat down lower and lower
in the bottom of the nest, and the egg, forced up,
would finally roll right into the cavity in his back—
the most irritable part of all. Whenever this occurred,
a sudden change that was like a fit would seize the
bird; he would stiffen, rise in the nest, his flabby

muscles made rigid, and stand erect, his back in a horizontal position, the head hanging down, the little naked wings held up over the back. In that position he looked an ugly, lumpish negro mannikin, standing on thinnest dwarf legs, his back bent, and elbows stuck up above the hollow flat back.

Once up on his small stiffened legs he would move backwards, firmly grasping the hairs and hair-like fibres of the nest-lining, and never swerving, until the rim of the cup-like structure was reached; and then standing, with feet sometimes below and in some cases on the rim, he would jerk his body, throwing the egg off or causing it to roll off. After that he would fall back into the nest and lie quite exhausted for some time, his jelly-like body rising and falling with his breathing.

These changes in the bird strongly reminded me of a person with an epileptic fit, as I had been accustomed to see it on the pampas, where, among the gauchos, epilepsy is one of the commonest maladies; —the sudden rigidity of muscle in some weak, sickly, flabby-looking person, the powerful grip of the hand, the strength in struggling, exceeding that of a man in perfect health, and finally, when this state is over, the weakness of complete exhaustion.

I witnessed several struggles with the egg, but at last, in spite of my watchfulness, I did not see it ejected. On returning after a very short absence, I found the egg had been thrown out and had rolled

down the bank, a distance of fourteen inches from
the nest.

The young cuckoo appeared to rest more quietly in
the nest now, but after a couple of hours the old
fidgeting began again, and increased until he was in
the same restless state as before. The rapid growth
of the birds made the position more and more miserable for the cuckoo, since the robin, thrust against
the side of the nest, would throw his head and neck
across the cuckoo's back, and he could not endure
being touched there. And now a fresh succession of
struggles began, the whole process being just the same
as when the egg was struggled with. But it was not
so easy with the young bird, not because of its greater
weight, but because it did not roll like the egg and
settle in the middle of the back ; it would fall partly
on to the cuckoo's back and then slip off into the nest
again. But success came at last, after many failures.
The robin was lying partly across the cuckoo's neck,
when, in moving its head, its little curved beak came
down and rested on the very centre of that irritable
hollow in the back of its foster-brother. Instantly the
cuckoo pressed down into the nest, shrinking away
as if hot needles had pricked him, as far as possible
from the side where the robin was lying against him,
and this movement of course brought the robin more
and more over him, until he was thrown right upon
the cuckoo's back.

Instantly the rigid fit came on, and up rose the

cuckoo, as if the robin weighed no more than a feather on him; and away backwards he went, right up the nest, without a pause, and standing actually on the rim, jerked his body, causing the robin to fall off, clean away from the nest. It fell, in fact, on to a large dock leaf five inches below the rim of the nest, and rested there.

After getting rid of his burden the cuckoo continued in the same position, perfectly rigid, for a space of five or six seconds, during which it again and again violently jerked its body, as if it had the feeling of the burden on it still. Then, the fit over, it fell back, exhausted as usual.

I had been singularly fortunate in witnessing the last scene and conclusion of this little bloodless tragedy in a bird's nest, with callow nestlings for *dramatis personæ*, this innocent crime and wrong, which is not a wrong since the cuckoo doesn't think it one. It is a little curious to reflect that a similar act takes place annually in tens of thousands of small birds' nests all over the country, and that it is so rarely witnessed.

Marvellous as the power of the young cuckoo is when the fit is on him, it is of course limited, and when watching his actions I concluded that it would be impossible for him to eject eggs and nestlings from any thrush's nest. The blackbird's would be too deep, and as to the throstle's, he could not move backwards up the sides of the cup-like cavity on account of the smooth plastered surface.

After having seen the young robin cast out I still

Cuckoo Ejecting Robin from the Nest

refrained from touching the nest, as there were yet
other things to observe. One was the presence, very
close to the nest, of the ejected nestling—what would
the parents do in the case? Before dealing with that
matter I shall conclude the history of the young cuckoo.
Having got the nest to himself he rested very
quietly, and it was not till the following day (July 1st)
that I allowed myself to touch him. He was, I found,
still irritable, and when I put back the eggs he had
thrown out he was again miserable in the nest, and
the struggle with the eggs was renewed until he got
rid of them as before. The next day the irritability
had almost gone, and in the afternoon he suffered
an egg or a pebble to remain in the nest with him
without jerking and wriggling about, and he made
no further attempt to eject it. This observation—
the loss of irritability on the fifth day after hatching
—agrees with that of Mr. Craig, whose account was
printed in the *Feathered World*, July 14, 1899.

The young cuckoo grew rapidly and soon trod his
nest into a broad platform, on which he reposed, a
conspicuous object in the scanty herbage on the bank.
We often visited and fed him, when he would puff
up his plumage and strike savagely at our hands, but
at the same time he would always gobble down the
food we offered. In seventeen days after being hatched
he left the nest and took up his position in an oak
tree growing on the bank, and there the robins continued feeding him for the next three days, after
which we saw no more of him.

I may add that in May 1901 a pair of robins built on the bank close to where the nest had been made the previous year, and that in this nest a cuckoo was also reared. The bird, when first seen, was apparently about four or five days old, and it had the nest to itself. Three ejected robin's eggs were lying on the bank a little lower down.

It is hardly to be doubted that the robins were the same birds that had reared the cuckoo in the previous season; and it is highly probable that the same cuckoo had returned to place her egg in their nest.

The end of the little history—the fate of the ejected nestling and the attitude of the parent robins—remains to be told. When the young cuckoo throws out the nestlings from nests in trees, hedges, bushes, and reeds, the victims, as a rule, fall some distance to the ground, or in the water, and are no more seen by the old birds. Here the young robin, when ejected, fell a distance of but five or six inches, and rested on a broad, bright green leaf, where it was an exceedingly conspicuous object; and when the mother robin was on the nest—and at this stage she was on it a greater part of the time—warming that black-skinned, toad - like, spurious babe of hers, her bright, intelli--gent eyes were looking full at the other one, just beneath her, which she had grown in her body and had hatched with her warmth, and was her very own. I watched her for hours; watched her when warming the cuckoo, when she left the nest and when she

returned with food, and warmed it again, and never
once did she pay the least attention to the outcast
lying there so close to her.   There, on its green leaf,
it remained, growing colder by degrees, hour by hour,
motionless, except when it lifted its head as if to
receive food, then dropped it again, and when, at
intervals, it twitched its body as if trying to move.
During the evening even these slight motions ceased,
though that feeblest flame of life was not yet extin-
guished; but in the morning it was dead and cold and
stiff; and just above it, her bright eyes on it, the mother
robin sat on the nest as before, warming her cuckoo.

How amazing and almost incredible it seems that
a being such as a robin, intelligent above most birds
as we are apt to think, should prove in this instance
to be a mere automaton!   The case would, I think,
have been different if the ejected one had made a
sound, since there is nothing which more excites the
parent bird, or which is more instantly responded to,
than the cry of hunger or distress of the young.   But
at this early stage the nestling is voiceless—another
point in favour of the parasite.   The sight of its young,
we see, slowly and dumbly dying, touches no chord
in the parent: there is, in fact, no recognition; once
out of the nest it is no more than a coloured leaf, or a
bird-shaped pebble, or fragment of clay.

It happened that my young fellow-watchers, seeing
that the ejected robin if left there would inevitably
perish, proposed to take it in to feed and rear it—

to *save* it, as they said; but I advised them not to attempt such a thing, but rather to *spare* the bird. To spare it the misery they would inflict on it by attempting to fill its parents' place. They had, so far, never kept a caged bird, nor a pet bird, and had no desire to keep one; all they desired to do in this case was to save the little outcast from death—to rear it till it was able to fly away and take care of itself. That was a difficult, a well-nigh impossible task. The bird, at this early stage, required to be fed at short intervals for about sixteen hours each day on a peculiar kind of food, suited to its delicate stomach—chiefly small caterpillars found in the herbage; and it also needed a sufficient amount by day and night of that animal warmth which only the parent bird could properly supply. They, not being robins, would give it unsuitable food, feed it at improper times, and not keep it at the right temperature, with the almost certain result that after lingering a few days it would die in their hands. But if by giving a great deal of time and much care they should succeed in rearing it, their foundling would start his independent life so handicapped, weakened in constitution by an indoor artificial bringing up, without the training which all young birds receive from their parents after quitting the nest, that it would be impossible for him to save himself. If by chance he should survive until August, he would then be set upon and killed by one of the adult robins already in possession of the ground. Now, when a bird at maturity perishes, it suffers in dying—sometimes very acutely; but if left to grow cold and fade out of life at this stage it can hardly be said to suffer. It is no more conscious than a chick in the shell; take from it the warmth that keeps it in being, and it drops back into nothingness without knowing and, we may say, without feeling anything. There may indeed be an incipient consciousness in that small, soft brain in its early vegetative stage, a first faint glimmer of a bright light to be, and a slight sensation of numbness may be actually felt as the body grows cold, but that would be all.

Pain is so common in the world; and, owing to the softness and sensitiveness induced in us by an indoor artificial life—since that softness of our bodies reacts on our minds—we have come to a false or an exaggerated idea of its importance, its *painfulness*, to put it that way; and we should therefore be but making matters worse, or rather making ourselves more miserable, by looking for and finding it where it does not exist.

The power to feel pain in any great degree comes into the bird's life after this transitional period, and is greatest at maturity, when consciousness and all the mental faculties are fully developed, particularly the passion of fear, which plays continually on the strings of the wild creature's heart with an ever varying touch, producing the feeling in all degrees from the slight disquiet, which is no sooner come than gone, to extremities of agonising terror. It would perhaps have a wholesome effect on their young minds, and save them from grieving overmuch at the death of a newly-hatched robin, if they would consider this fact of the pain that is and must be. Not the whole subject—the fact that as things are designed in this world of sentient life there can be no good, no sweetness or pleasure in life, nor peace and contentment and safety, nor happiness and joy, nor any beauty or strength or lustre, nor any bright and shining quality of body or mind, without pain, which is not an accident nor an incident, nor something ancillary to life, but is involved in and a part of life, of its very colour and texture. That would be too long to speak about; all I meant was to consider that small part of the fact, the necessary pain to and destruction of the bird life around them and in the country generally.

Here, for instance, without going farther than a hundred yards from the house in any direction, they could put their hands in nests in trees and bushes, and on the ground, and in the ivy, and in the old outhouses, and handle and count about one hundred and thirty young birds not yet able to fly. Probably more than twice that number would be successfully reared during the season. How many, then, would be reared in the whole parish! How many in the entire New Forest district, in the whole county of Hampshire, in the entire kingdom! Yet when summer came round again they would find no more birds than they had now. And so it would be in all places; all that incalculable increase would have perished. Many millions would be devoured by rapacious birds and beasts; millions more would perish of hunger and cold; millions of migrants would fall by the way, some in the sea and some on land; those that returned from distant regions would be but a remnant, and the residents that survived through the winter, these, too, would be nothing but a remnant. It is not only that this inconceivable amount of bird life must be destroyed each year, but we cannot suppose that death is not a painful process. In a vast majority of cases, whether the bird slowly perishes of hunger and weakness, or is pursued and captured by birds and beasts of prey, or is driven by cold adverse winds and storms into the waves, the pain, the agony must be great. The least painful death is undoubtedly that of the bird that, weakened by want of sustenance, dies by night of cold in severe weather. It is indeed most like the death of the nestling, but a few hours out of the shell, which has been thrown out of the nest, and which soon grows cold, and dozes its feeble, unconscious life away.

We may say, then, that of all the thousand forms of death which Nature has invented to keep her too rapidly multiplying creatures within bounds, that which is brought about by the singular instinct of the young cuckoo in the nest is the most merciful or the least painful.

I am not sure that I said all this, or marshalled fact and argument in the precise order in which they are here set down. I fancy not, as it seems more than could well have been spoken while we

were standing there in the late evening sunlight by
that primrose bank, looking down on the little
flesh - coloured mite in its scant clothing of black
down, fading out of life on its cold green leaf. But
what was said did not fail of its effect, so that my
young tender - hearted hearers, who had begun to
listen with moist eyes, secretly accusing me perhaps of
want of feeling, were content in the end to let it be—
to go away and leave it to its fate in that mysterious
green world we, too, live in and do not understand, in
which life and death and pleasure and pain are inter-
woven light and shade.

at Boldre.

CHAPTER II

BETWEEN the Boldre and the Exe, or Beaulieu river,
there is a stretch of country in most part flat and
featureless. It is one of those parts of the Forest
which have a bare and desolate aspect; here in
places you can go a mile and not find a tree or
bush, where nothing grows but a starved-looking
heath, scarcely ankle-deep. Wild life in such places
is represented by a few meadow pipits and small
lizards. There is no doubt that this barrenness and
naked appearance is the result of the perpetual cutting of heath and gorse, and the removal of the thin
surface soil for fuel.

Those who do not know the New Forest, or know
it only as a collecting or happy hunting-ground of
eggers and "lepidopterists," or as artists in search of
paintable woodland scenery know it, and others who

make it a summer holiday resort, may say that this
abuse is one which might and should be remedied.
They would be mistaken.   What I and a few others
who use their senses see and hear in this or that spot,
is, in every case, a very small matter, a visible but
an infinitesimal part of that abuse of the New Forest
which is old and chronic, and operates always, and
is common to the whole area, and, as things are, irre-
mediable.   To discover and denounce certain things
which ought not to be, to rail against Verderers, who
are after all what they cannot help being, is about as
profitable as it would be to " damn the nature of things."

It must be borne in mind that the Forest area has
a considerable population composed of commoners,
squatters, private owners, who have inherited or pur-
chased lands originally filched from the Forest; and
of a large number of persons who reside mostly in
the villages, and are private residents, publicans, shop-
keepers, and lodging-house keepers.   All these people
have one object in common—to get as much as they
can out of the Forest.   It is true that a large proportion of them, especially those who live in the villages,
which are now rapidly increasing their populations,
are supposed not to have any Forest rights; but they
do as a fact get something out of it; and we may
say that, generally, all the people in the Forest dine
at one table, and all get a helping out of most of the
dishes going, though the first and biggest helpings
are for the favoured guests.

Those who have inherited rights have indeed come to look on the Forest as in a sense their property. What is given or handed over to them is not in their view their proper share: they take this openly, and get the balance the best way they can—in the dark generally. It is not dishonest to help yourself to what belongs to you; and they must live—must have their whack. They have, in fact, their own moral code, their New Forest conscience, just as other men—miners, labourers on the land, tradesmen, gamekeepers, members of the Stock Exchange, for instance—have each their corporate code and conscience. It may not be the general or the ideal or speculative conscience, but it is what may be called their working conscience. One proof that much goes on in the dark, or that much is winked at, is the paucity of all wild life which is worth any man's while to take in a district where pretty well everything is protected on paper. Game, furred and feathered, would not exist at all but for the private estates scattered through the Forest, in which game is preserved, and from which the depleted Forest lands are constantly being restocked. Again, in all this most favourable country no rare or beautiful species may be found: it would be safer for the hobby, the golden oriole, the hoopoe, the harrier, to nest in a metropolitan park than in the loneliest wood between the Avon and Southampton Water. To introduce any new species, from the biggest—the capercailzie and the great bustard —to the smallest quail, or any small passerine bird with a spot of brilliant colour on its plumage, would be impossible.

The New Forest people are, in fact, just what circumstances have made them. Like all organised beings, they are the creatures of, and subject to, the conditions they exist in; and they cannot be other than they are —namely, parasites on the Forest. And, what is more, they cannot be educated, or preached, or worried out of their ingrained parasitical habits and ways of thought. They have had centuries—long centuries—of practice to make them cunning, and the effect of more stringent regulations than those now in use would only be to polish and put a better edge on that weapon which Nature has given them to fight with.

This being the conclusion, namely, that "things are what they are, and the consequences of them will be what they will be," some of my readers, especially those in the New Forest, may ask, Why, then, say anything about it? why not follow the others who have written books and books and books about the New Forest, books big and books little, from Wise, his classic, and the Victoria History, down to the long row of little rosy guide-books? They saw nothing of all this; or if they saw unpleasant things they thought it better to hold their tongues, or pens, than to make people uncomfortable.

I confess it would be a mistake, a mere waste of words, to bring these hidden things to light if it could be believed that the New Forest, in its condition and management, will continue for any length of time to be what it is and has been—just that and nothing more. A district in England, it is true, but out of the way, remote, a spot to be visited once or twice in a lifetime just to look at the scenery, like Lundy or the Scilly Isles or the Orkneys. But it cannot be believed. The place itself, its curious tangle of ownership; government by and rights of the crown, of private owners, commoners, and the public, is what it has always been; but many persons have now come to think and to believe that the time is approaching when there will be a disentanglement and a change.

The Forest has been known and loved by a limited number of persons always; the general public have only discovered it in recent years. For one visitor twenty years ago there are scores, probably hundreds, to-day. And year by year, as motoring becomes more common, and as cycling from being general grows, as it will, to be universal, the flow of visitors to the Forest will go on at an ever-increasing rate, and the hundreds of to-day will be thousands in five years' time. With these modern means of locomotion, there is no more attractive spot than this hundred and fifty square miles of level country which contains the most beautiful forest scenery in England. And as it grows in favour in all the country as a place of recreation and refreshment, the subject of its condition and management, and the ways of its inhabitants, will receive an increased attention. The desire will grow that it shall not be spoilt, either by the

authorities or the residents, that it shall not be turned
into townships and plantations, nor be starved, nor its
wild life left to be taken and destroyed by any one and
every one.   It will be seen that the "rights" I have
spoken of, with the unwritten laws and customs which
are kept more or less in the dark, are in conflict with
the better and infinitely more important rights of the
people generally—of the whole nation.   Once all this
becomes common knowledge, that which some now
regard as a mere dream, a faint hope, something too
remote for us to concern ourselves about, will all at
once appear to us as a practical object—something to
be won by fighting, and certainly worth fighting for.

It may be said at once, and I fancy that any one who
knows the inner life of the Forest people will agree with
me, that so long as these are in possession (and here all
private owners are included) there can be no great
change, no permanent improvement made in the Forest.
That is the difficulty, but it is not an insuperable one.
Public opinion, and the desire of the people for anything, is a considerable force to-day; so that, inspired
by it, the most timid and conservative governments
are apt all at once to acquire an extraordinary courage.
Sustained by that outside force, the most tender-
hearted and sensitive Prime Minister would not in the
least mind if some persons were to dub him a second
and worse William the Bastard.

The people in this district have a curious experiment
to show the wonderful power of the Forest fly in retain-

ing its grasp. A man takes the fly between his finger and thumb, and with the other hand holds a single hair of a cow or horse for it to seize, then gently pulls hair and fly apart. The fly does not release his hold— he splits the hair, or at any rate shaves a piece off right down to the fine end with his sharp, grasping claw. Doubtless the human parasite will, when his time comes, show an equal tenacity; he will embrace the biggest and oldest oak he knows, and to pluck him from his beloved soil it will be necessary to pull up the tree by its roots. But this is a detail, and may be left to the engineers.

Beyond that starved, melancholy wilderness, the sight of which has led me into so long a digression, one comes to a point which overlooks the valley of the Exe; and here one pauses long before going down to the half-hidden village by the river. Especially if it is in May or June, when the oak is in its "glad light grene," for that is the most vivid and beautiful of all vegetable greens, and the prospect is the greenest and most soul-refreshing to be found in England. The valley is all wooded and the wood is all oak—a continuous oak-wood stretching away on the right, mile on mile, to the sea. The sensation experienced at the sight of this prospect is like that of the traveller in a dry desert when he comes to a clear running stream and drinks his fill of water and is refreshed. The river is tidal, and at the full of the tide in its widest part beside the village its appearance is of a small inland lake, grown round with oaks—old trees that stretch their horizontal branches far out and wet their lower leaves in the salt water. The village itself that has this setting, with its ancient water-mill, its palace of the Montagus, and the Abbey of Beaulieu, a grey ivied ruin, has a distinction

BEAULIEU MILL

above all Hampshire villages, and is unlike all others in its austere beauty and atmosphere of old-world seclusion and quietude. Above all, is that quality which the mind imparts — the expression due to romantic historical associations.

One very still, warm summer afternoon I stood on the margin, looking across the sheet of glassy water at a heron on the farther side, standing knee-deep in the shallow water patiently watching for a fish, his grey figure showing distinctly against a background of bright green sedges. Between me and the heron scores of swallows and martins were hawking for flies, gliding hither and thither a little above the glassy surface, and occasionally dropping down to dip and wet their under plumage in the water. And all at once, fifty yards out from the margin, there was a great splash, as if a big stone had been flung out into the lake; and then two or three moments later out from the falling spray and rocking water rose a swallow, struggling laboriously up, its plumage drenched, and flew slowly away. A big pike had dashed at and tried to seize it at the moment of dipping in the water, and the swallow had escaped as by a miracle. I turned round to see if any person was near, who might by chance have witnessed so strange a thing, in order to speak to him about it. There was no person within sight, but if on turning round my eyes had encountered the form of a Cistercian monk, returning from his day's labour in the fields, in his dirty black-and-white robe, his implements on his shoulders, his face and hands begrimed with dust and sweat, the apparition on that day, in the mood I was in, would not have greatly surprised me.

The atmosphere, the expression of the past may so attune the mind as almost to produce the illusion that the past is now.

But more than old memories, great as their power

over the mind is at certain impressible moments, and more than Beaulieu as a place where men dwell, is that ineffable freshness of nature, that verdure that like the sunlight and the warmth of the sun penetrates to the inmost being.   Here I have remembered the old ornithologist Willughby's suggestion, which no longer seemed fantastic, that the furred and feathered creatures inhabiting arctic regions have grown white by force of imagination and the constant intuition of snow.   And here too I have recalled that modern fancy that the soul in man has its proper shape and colour, and have thought that if I came hither with a grey or blue or orange or brown soul, its colour had now changed to green.   The pleasure of it has detained me long days in spring, often straying by the river at its full, among the broadly-branching oaks, delighting my sight with the new leaves

> against the sun shene,
> Some very red, and some a glad light grene.

Yet these same oak woods, great as their charm is, their green everlasting gladness, have a less enduring hold on the spirit than the open heath, though this may look melancholy and almost desolate on coming to it from those sunlit emerald glades with a green thought in the soul.   It seems enough that it is open, where the wind blows free, and there is nothing between us and the sun.   It is a passion, an old ineradicable instinct in us: the strongest impulse in children, savage or civilised, is to go out into some open place.   If a man be capable of an exalted mood, of a sense of absolute freedom, so that he is no longer flesh and spirit but both in one, and one with nature, it comes to him like some miraculous gift on a hill or down or wide open heath. " You never enjoy the earth aright," wrote Thomas Traherne in his *Divine Raptures*, " until the sun itself floweth through your veins, till you are clothed with the heavens and crowned with the stars, and perceive yourself to be the sole heir of the whole world."

It may be observed that we must be out and well away from the woods and have a wide horizon all around in order to feel the sun flowing through us. Many of us have experienced these "divine raptures," this sublimated state of feeling; and such moments are perhaps the best in our earthly lives; but it is mainly the Trahernes, the Silurist Vaughans, the Newmans, the Frederic Myers, the Coventry Patmores, the Wordsworths, that speak of them, since such moods best fit, or can be made to fit in with their philosophy, or mysticism, and are, to them, its best justification.

This wide heath, east of Beaulieu, stretching miles away towards Southampton Water, looks level to the eye. But it is not so; it is grooved with long valley-like depressions with marshy or boggy bottoms, all draining into small tributaries of the Dark Water, which flows into the Solent near Lepe. In these bottoms and in all the wet places the heather and furze mixes with or gives place to the bog myrtle, or golden withy; and on the spongiest spots the fragrant yellow stars of the bog asphodel are common in June. These spots are exceedingly rich in colour, with greys and emerald greens and orange yellows of moss and lichen, flecked with the snow-white of cotton-grass.

Here, then, besides that cause of contentment which we find in openness, there is fragrance in fuller measure than in most places. One may wade through acres of myrtle, until that subtle delightful odour is in one's skin and clothes, and in the air one breathes, and seems at last to penetrate and saturate the whole being, and smell seems to be for a time the most important of the senses.

Among the interesting birds that breed on the heath, the nightjar is one of the commonest. A keen naturalist, Mr. E. A. Bankes, who lived close by, told me that he had marked the spot where he had found a pair of young birds, and that each time he rode over the heath he had a look at them, and as they remained there until able to fly, he concluded that it is not true that the parent birds remove the young when the nest has been discovered.

I was not convinced, as it did not appear that he had handled the young birds: he had only looked at them while sitting on his horse. The following summer I found a pair of young not far from the same spot: they were half-fledged and very active, running into the heath and trying to hide from me, but I caught and handled them for some minutes, the parent bird remaining near, uttering her cries. I marked the spot and went back next day, only to find that the birds had vanished.

The snipe, too, is an annual breeder, and from what I saw of it on the heath I think we have yet something to learn concerning the breeding habits of that much-observed bird. The parent bird is not so wise as most mothers in the feathered world, since her startling cry of alarm, sounding in a small way like the snort of a frightened horse, will attract a person to the spot where she is sheltering her young among the myrtle. She will repeat the cry at intervals a dozen times without stirring or attempting to conceal the young. But she does not always act in the same way. Sometimes she has risen to a great height and begun circling above me, the circles growing smaller or larger as I came nearer or went farther from the spot where the young were lurking.

It was until recently a moot question as to whether or not the female snipe made the drumming or bleating sound; some of the authorities say that this sound proceeds only from the male bird. I have no doubt that both birds make the sound. Invariably when I disturbed a snipe with young, and when she mounted high in the air, to wheel round and round, uttering her anxious cries, she dashed downwards at intervals, and produced the bleating or drumming which the male birds emit when playing about the sky.

In all cases where I have found young snipe there was but one old bird, the female, no doubt. In some instances I have spent an hour with the young birds by me, or in my hands, waiting for the other parent to appear; and I am almost convinced that the care of the young falls wholely on the female.

The redshank, that graceful bird with a beautiful voice, breeds here in most years, and is in a perpetual state of anxiety so long as a human figure remains in sight. A little while ago the small vari-coloured stonechat or fuzz-jack, with red breast, black head and white collar, sitting upright and motionless, like a painted image of a bird, on the topmast' spray of a furze bush, then flitting to perch on another bush, then to another; for ever emitting those two little contrasted sounds — the gutteral chat and the clear, fretful pipe — had seemed to me the most troubled and full of care and worries of all Nature's feathered children—so sorrowful, in spite of his pretty harlequin dress! Now his trouble seems a small thing, and not to be regarded in the presence of the larger, louder redshank. As I walk he rises a long way ahead, and wheeling about comes towards me—he and she, and by-and-by a second pair, and perhaps a third; they come with measured pulsation of the long, sharp, white-banded wings; and the first comer sweeps by and returns again to meet the others, clamouring all the time, calling on them to join in the outcry until the whole air seems full of their trouble. To and fro he flies, to this side and that; and finally, as if in imitation of the small, fretful stonechat, he sweeps down to alight on the topmost spray of some small tree or tall bush—not a furze but a willow; and as it is an insecure stand for a bird of his long thin wading legs, he stands lightly, balancing himself with his wings; beautiful in his white and pale-grey plumage, and his slender form, on that airy perch of the willow in its grey-green leaves and snow-white catkins; and balanced there, he still continues his sorrowful anxious cries— ever crying for me to go—to go away and leave him in peace. I leave him reluctantly, and have my reward, for no sooner does he see me going than his anxious cries change to that beautiful wild pipe, un- rivalled except by the curlew among shore birds.

Worst of all birds that can have no peace in their lives so long as you are in sight is the peewit. The harsh wailing sound of his crying voice as he wheels about overhead, the mad downward rushes, when his wings creak as he nears you, give the idea that he is almost crazed with anxiety; and one feels ashamed at causing so much misery. Oh, poor bird! is there no way to make you understand without leaving the ground, that your black-spotted, olive-coloured eggs are perfectly safe; that a man can walk about on the heath and be no more harmful to you than the Forest ponies, and the ragged donkey browsing on a furze bush, and the cow with her tinkling bell? I stand motionless, looking the other way; I sit down to think; I lie flat on my back with hands clasped behind my head, and gaze at the sky, and still the trouble goes on—he will not believe in me, nor tolerate me. There is nothing to do but get up and go away out of sight and sound of the peewits.

It appears to me that this sympathy for the lower animals is very much a matter of association—an overflow of that regard for the rights of and compassion for others of our kind which are at the foundations of the social instinct. The bird is a red- and a warm-blooded being—we have seen that its blood is red, and when we take a living bird in our hands we feel its warmth and the throbbing of its breast: therefore birds are related to us, and with that red human blood they have human passions. Witness the peewit —the mother bird, when you have discovered or have come near her downy little one—could any human mother, torn with the fear of losing her babe, show her unquiet and disturbed state in a plainer, more understandable way! But in the case of creatures of another division in the kingdom of life—non-vertebrates, without sensible heat, and with a thin colourless fluid instead of red blood, as if like plants they had only a vegetative life—this sympathy is not felt as a rule. When, in some exceptional case, the feeling is there, it is because some human association has come into the mind in spite of the differences between insect and man.

Walking on this heath I saw a common green grass-hopper, disturbed at my step, leap away, and by chance land in a geometric web in a small furze bush. Caught in the web, it began kicking with its long hind legs, and would in three seconds have made its escape. But mark what happened. Directly over the web, and above the kicking grasshopper, there was a small, web-made, thimble-shaped shelter, mouth down, fastened to a spray, and the spider was sitting in it. And looking down it must have seen and known that the grasshopper was far too big and strong to be held in that frailest snare, that it would be gone in a moment and the net torn to pieces. It also must have seen and known that it was no wasp nor dangerous insect of any kind; and so, instantly, straight and swift as a leaden plummet, it dropped out of the silvery bell it lived in on to the grasshopper and attacked it at the head. The falces were probably thrust into the body between the head and prothorax, for almost instantly the struggle ceased, and in less than three seconds the victim appeared perfectly dead.

What interested me in this sight was the spider, an Epeira of a species I had never closely looked at before, a little less in size than our famous Epeira diadema—our common garden spider, with the pretty white diadem on its velvety, brown abdomen. This heath spider was creamy-white in colour, the white deepening to warm buff all round at the sides, and to a deeper tint on the under surface. It was curiously and prettily coloured; and, being new to me, its image was vividly impressed on my mind.

As to what had happened, that did not impress me at all. I could not, like the late noble poet who cherished an extreme animosity against the spider, and inveighed against it in brilliant, inspired verse, remember and brood sadly on the thought of the fairy forms that are .its victims—

> The lovely births that winnow by,
> Twin-sisters of the rainbow sky :
> Elf-darlings, fluffy, bee-bright things,
> And owl-white moths with mealy wings.

Nor could I, like him, break the creature's toils, nor take the dead from its gibbet, nor slay it on account of its desperate wickedness. These are mere house-bred feelings and fancies, perhaps morbid; he who walks out-of-doors with Nature, who sees life and death as sunlight and shadow, on witnessing such an incident wishes the captor a good appetite, and, passing on, thinks no more about it. For any day in summer, sitting by the water, or in a wood, or on the open heath, I note little incidents of this kind; they are always going on in thousands all about us, and one with trained eye cannot but see them; but no feeling is excited, no sympathy, and they are no sooner seen than forgotten. But, as I said, there are exceptional cases, and here is one which refers to an even more insignificant creature than a field grasshopper—a small dipterous insect—and yet I was strangely moved by it.

The insect was flying rather slowly by me over the heath—a thin, yellow-bodied, long-legged creature, a Tipula, about half as big as our familiar crane-fly. Now, as it flew by me about on a level with my thighs, up from the heath at my feet shot out a second insect, about the same size as the first, also a Dipteron, but of another family—one of the Asilidæ, which are rapacious. The Asilus was also very long-legged, and seizing the other with its legs, the two fell together to the ground. Stooping down, I witnessed the struggle. They were locked together, and I saw the attacking insect raise his head and the forepart of his body so as to strike, then plunge his rostrum like a dagger in the soft part of his victim's body. Again and again he raised and buried his weapon in the other, and the other still refused to die or to cease struggling. And this little fight and struggle of two flies curiously moved me, and for some time I could not get over the feeling of intense repugnance it excited. This feeling was wholly due to association : the dagger-like weapon and the action of the insect were curiously human-like, and I had seen just such a combat between two men, one fallen and the other on him, raising and striking down with his knife. Had I never witnessed such an incident, the two flies struggling, one killing the other, would have produced

D

no such feeling, and would not have been re-membered.

We live in thoughts and feelings, not in days and years—

In feelings, not in figures on a dial,

as some poet has said, and, recalling an afternoon and an evening spent on this heath, it does not seem to my mind like an evening passed alone in a vacant place, in the usual way, watching and listening and think-ing of nothing, but an eventful period, which deeply moved me, and left an enduring memory.

The sun went down, and though the distressed birds had cried till they were weary of crying, I did not go away. Something on this occasion kept me, in spite of the gathering gloom and a cold wind—bit-terly cold for June—which blew over the wide heath. Here and there the rays from the setting sun fell upon and lit up the few mounds that rise like little islands out of the desolate brown waste. These are the Pixie mounds, the barrows raised by probably pre-historic men, a people inconceivably remote in time and spirit from us, whose memory is pale in our civilised days.

There are times and moods in which it is revealed to us, or to a few among us, that we are a survival of the past, a dying remnant of a vanished people, and are like strangers and captives among those who do not understand us, and have no wish to do so;

THE BARROW ON THE HEATH

whose language and customs and thoughts are not ours.
That "world-strangeness," which William Watson and
his fellow-poets prattle in rhyme about, those, at all
events, who have what they call the "note of moder-
nity" in their pipings, is not in me as in them. The
blue sky, the brown soil beneath, the grass, the trees,
the animals, the wind, and rain, and sun, and stars are
never strange to me; for I am in and of and am one
with them; and my flesh and the soil are one, and
the heat in my blood and in the sunshine are one,
and the winds and tempests and my passions are one.
I feel the "strangeness" only with regard to my fellow-
men, especially in towns, where they exist in conditions
unnatural to me, but congenial to them; where they
are seen in numbers and in crowds, in streets and
houses, and in all places where they gather together;
when I look at them, their pale civilised faces, their
clothes, and hear them eagerly talking about things
that do not concern me. They are out of my world
—the real world. All that they value, and seek and
strain after all their lives long, their works and sports
and pleasures, are the merest baubles and childish
things; and their ideals are all false, and nothing
but by-products, or growths, of the artificial life—
little funguses cultivated in heated cellars.

In such moments we sometimes feel a kinship with,
and are strangely drawn to, the dead, who were not
as these; the long, long dead, the men who knew
not life in towns, and felt no strangeness in sun and

wind and rain. In such a mood on that evening I
went to one of those lonely barrows; one that rises to
a height of nine or ten feet above the level heath,
and is about fifty yards round. It is a garden in the
brown desert, covered over with a dense growth of
furze bushes, still in flower, mixed with bramble and
elder and thorn, and heather in great clumps, blooming, too, a month before its time, the fiery purple-
red of its massed blossoms, and of a few tall, tapering spikes of foxglove, shining against the vivid green
of the young bracken.

All this rich wild vegetation on that lonely mound
on the brown heath!

Here, sheltered by the bushes, I sat and saw the
sun go down, and the long twilight deepen till the
oak woods of Beaulieu in the west looked black on
the horizon, and the stars came out: in spite of the
cold wind that made me shiver in my thin clothes, I
sat there for hours, held by the silence and solitariness
of that mound of the ancient dead.

Sitting there, profoundly sad for no apparent cause,
with no conscious thought in my mind, it suddenly
occurred to me that I knew that spot from of old,
that in long past forgotten years I had often come
there of an evening and sat through the twilight,
in love with the loneliness and peace, wishing that
it might be my last resting-place. To sleep there
for ever—the sleep that knows no waking! We say
it, but do not mean—do not believe it. Dreams do

come to give us pause; and we know that we have
lived. To dwell alone, then, with this memory of life
in such a spot for all time! There are moments in
which the thought of death steals upon and takes us
as it were by surprise, and it is then exceeding bitter.
It was as if that cold wind blowing over and making
strange whispers in the heather had brought a sudden
tempest of icy rain to wet and chill me.

This miserable sensation soon passed away, and,
with quieted heart, I began to grow more and more
attracted by the thought of resting on so blessed a
spot. To have always about me that wildness which
I best loved—the rude incult heath, the beautiful
desolation; to have harsh furze and ling and bramble
and bracken to grow on me, and only wild creatures
for visitors and company. The little stonechat, the
tinkling meadow pipit, the excited whitethroat to
sing to me in summer; the deep-burrowing rabbit to
bring down his warmth and familar smell among my
bones; the heat-loving adder, rich in colour, to find
when summer is gone a dry safe shelter and hiber-
naculum in my empty skull.

So beautiful did the thought appear that I could
have laid down my life at that moment, in spite of
death's bitterness, if by so doing I could have had
my desire. But no such sweet and desirable a thing
could be given me by this strange people and race
that possess the earth, who are not like the people
here with me in the twilight on the heath. For I

thought, too, of those I should lie with, having with them my after life; and thinking of them I was no longer alone. I thought of them not as others think, those others of a strange race. What *do* they think? They think so many things! The materialist, the scientist, would say: They have no existence; they ceased to be anything when their flesh was turned to dust, or burned to ashes, and their minds, or souls, were changed to some other form of energy, or motion, or affection of matter, or whatever they call it. The believer would not say of them, or of the immaterial part of them, that they had gone into a world of light, that in a dream or vision he had seen them walking in an air of glory; but he might hold that they had been preached to in Hades some nineteen centuries ago, and had perhaps repented of their barbarous deeds. Or he might think, since he has considerable latitude allowed him on the point, that the imperishable parts of them are here at this very spot, tangled in dust that was once flesh and bones, sleeping like chrysalids through a long winter, to be raised again at the sound of a trumpet blown by an angel to a second conscious life, happy or miserable as may be willed.

I imagine none of these things, for they were with me in the twilight on the barrow in crowds, sitting and standing in groups, and many lying on their sides on the turf below, their heads resting in their hands. They, too, all had their faces turned towards

Beaulieu. Evening by evening for many and many a century they had looked to that point, towards the black wood on the horizon, where there were people and sounds of human life. Day by day for centuries they had listened with wonder and fear to the Abbey bells, and to the distant chanting of the monks. And the Abbey has been in ruins for centuries, open to the sky and overgrown with ivy; but still towards that point they look with apprehension, since men still dwell there, strangers to them, the little busy eager people, hateful in their artificial indoor lives, who do not know and who care nothing for them, who worship not and fear not the dead that are underground, but dig up their sacred places and scatter their bones and ashes, and despise and mock them because they are dead and powerless.

It is not strange that they fear and hate. I look at them—their dark, pale, furious faces—and think that if they could be visible thus in the daylight, all who came to that spot or passed near it would turn and fly with a terrifying image in their mind which would last to the end of life. But they do not resent my presence, and would not resent it were I permitted to come at last to dwell with them for ever. Perhaps they know me for one of their tribe,—know that what they feel I feel, would hate what they hate.

Has it not been said that love itself is an argument in favour of immortality? All love—the love of men and women, of a mother for her child, of a friend for a friend—the love that will cause him to lay down his life for another. Is it possible to believe, they say, that this beautiful sacred flame can be darkened for ever when soul and body fall asunder? But love without hate I do not know and cannot conceive; one implies the other. No good and no bad quality or principle can exist (for me) without its opposite. As old Langland wisely says—

> For by luthere men know the good;
> And whereby wiste men which were white
> If all things black were?

Beaulieu Abbey.

CHAPTER III

A favourite New Forest haunt—Summertide—Young blackbird's call—Abundance of blackbirds and thrushes and destruction of young—Starlings breeding—The good done by starlings—Perfume of the honeysuckle—Beauty of the hedge rose—Cult of the rose—Lesser whitethroat—His low song—Common and lesser whitethroat—In the woods—A sheet of bracken—Effect of broken surfaces—Roman mosaics at Silchester—Why mosaics give pleasure—Woodland birds—Sound of insect life—Abundance of flies—Sufferings of cattle—Dark Water—Biting and teasing flies—Feeding the fishes and fiddlers with flies.

LOOKING away from Beaulieu towards Southampton Water there is seen on the border of the wide brown heath a long line of tall firs, a vast dark grove forming the horizon on that side. This is the edge of an immense wood, and beyond the pines which grow by the heath, it is almost exclusively oak with an undergrowth of holly. It is low-lying ground with many streams and a good deal of bog, and owing to the dense undergrowth and the luxuriance of vegetation generally this part of the forest has a ruder, wilder appearance than at any other spot. Here, too, albeit the nobler bird and animal forms are absent, as is indeed the case in all the New Forest district, animal life generally is in greatest profusion and variety. This wood with its surrounding heaths, bogs, and farm lands, has been my favourite summer resort and hunting ground for some years past. With a farm-house not many minutes' walk from the forest for a home, I have here spent long weeks at a time, rambling in the woods every day and all day long, for the most time out of sight of human habitations, and always with the feeling that I was in my own territory, where everything was as Nature made it and as I liked it to be. Never once in all my rambles did I encounter that hated being, the collector, with his white, spectacled town face and green butterfly net. In this out-of-the-way corner of the Forest one could imagine the time come when this one small piece of England which lies between the Avon and Southampton Water will be a sanctuary for all rare and beautiful wild life and a place of refreshment to body and soul for all men.

The richest, fullest time of the year is when June is wearing to an end, when one knows without the almanac that spring is over and gone. Nowhere in England is one more sensible of the change to fullest summer than in this low-lying, warmest corner of Hampshire.

The cuckoo ceases to weary us with its incessant call, and the nightingale sings less and less frequently. The passionate season is well-nigh over for the birds; their fountain of music begins to run dry. The cornfields and waste grounds are everywhere splashed with the intense scarlet of poppies. Summer has no rain in all her wide, hot heavens to give to her thirsty fields, and has sprinkled them with the red fiery moisture from

her own veins.   And as colour changes, growing deeper
and more intense, so do sounds change: for the songs of
yesterday there are shrill hunger-cries.

One of the oftenest heard in all the open woods, in
hedges, and even out in the cornfields is the curious
musical call of the young blackbird.   It is like the
chuckle of the adult, but not so loud, full, happy, and
prolonged; it is shriller, and drops at the end to a
plaintive, impatient sound, a little pathetic—a cry of
the young bird to its too long absent mother.   When
very hungry he emits this shrill musical call at intervals
of ten to fifteen seconds; it may be heard distinctly a
couple of hundred yards away.

The numbers of young blackbirds and throstles
apparently just out of the nest astonish one.   They are
not only in the copses and hedges, and on almost every
roadside tree, but you constantly see them on the
ground in the lanes and public roads, standing still,
quite unconscious of danger.   The poor helpless bird
looks up at you in a sort of amazement, never having
seen men walking or riding on bicycles; but he
hesitates, not knowing whether to fly away or stand
still.   Thrush or blackbird, he is curiously interesting
to look at.   The young thrush, with his yellowish-white
spotty breast, the remains of down on his plumage, his
wide yellow mouth, and raised head with large, fixed,
toad-like eyes, has a distinctly reptilian appearance.
Not so the young blackbird, standing motionless on the
road, in doubt too as to what you are, his short tail raised,

giving him an incipient air of blackbird jauntiness; his plumage not brown, indeed, as we describe it, but rich chestnut black, like the chestnut-black hair of a beautiful Hampshire girl of that precious type with oval face

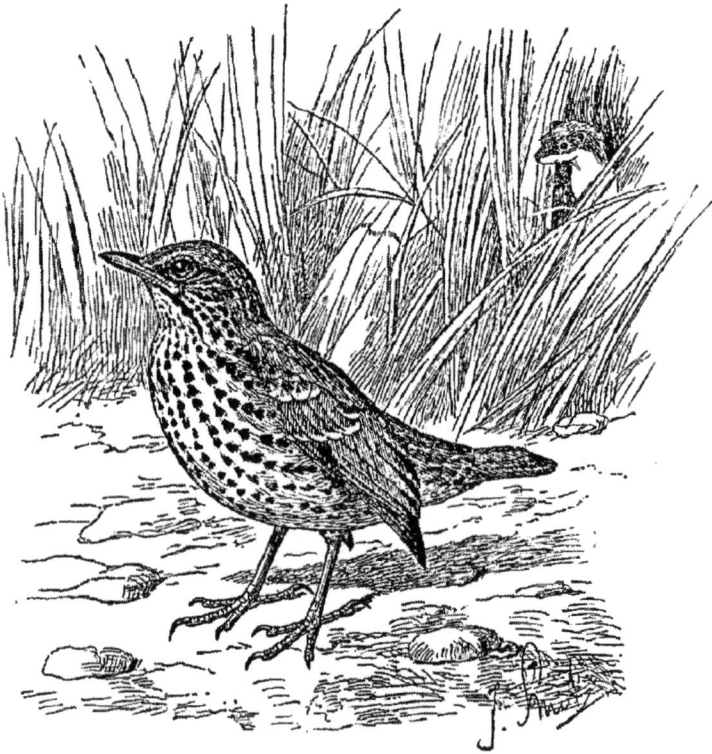

YOUNG THRUSH

and pale dark skin. A pretty creature, rich in colour, with a musical, pathetic voice, waiting so patiently to be visited and fed, and a weasel perhaps watching him from the roadside grass with hungry, bright little eyes!

How they die—thrushes and blackbirds—at this perilous period in their lives! I sometimes see what looks like a rudely-painted figure of a bird on the hard road: it is a young blackbird that had not the sense to get out of the way of a passing team, and was crushed flat by a hoof or wheel. It is but one in a thousand that perishes in that way. One has to remember that these two species of thrush—throstle and blackbird—are in extraordinary abundance, that next to starlings and chaffinches they abound over all species; that they are exceedingly prolific, beginning to lay in this southern county in February, and rearing at least three broods in the season; and that when winter comes round again the thrush and blackbird population will be just about what it was before.

Fruit-eating birds do not much vex the farmer in this almost fruitless country. Thrushes and finches and sparrows are nothing to him: the starling, if he pays any attention to the birds, he looks on as a good friend.

At the farm there are two very old yew trees growing in the back-yard, and one of these, in an advanced state of decay, is full of holes and cavities in its larger branches. Here about half-a-dozen pairs of starlings nest every year, and by the middle of June there are several broods of fully-fledged young. At this time it was amusing to watch the parent birds at their task, coming and going all day long, flying out and away straight as arrows to this side and that, every bird to

its own favourite hunting-ground. Some had their
grounds in the meadow, just before the house where
the cows and geese were, and it was easy to watch their
movements. Out of the yew the bird would shoot, and
in ten or twelve seconds would be down walking about
in that busy, plodding, rook-like way the starling has
when looking for something; and presently, darting
his beak into the turf, he would drag out something
large, and back he would fly to his young with a big,
conspicuous, white object in his beak. These white
objects which he was busily gathering every day, from
dawn to dark, were full-grown grubs of the cockchafer.
When watching these birds at their work it struck me
that the enormous increase of starlings all over the
country in recent years may account for the fact that
great cockchafer years do not now occur. In former
years these beetles were sometimes in such numbers
that they swarmed in the air in places, and stripped
the oaks of their leaves in midsummer. It is now more
than ten years since I saw cockchafers in considerable
numbers, and for a long time past I have not heard
of their appearance in swarms anywhere.

The starling is in some ways a bad bird, a cherry
thief, and a robber of other birds' nesting-places; yaffle
and nuthatch must hate him, but if his ministrations
have caused an increase of even one per cent. in the
hay crop, and the milk and butter supply, he is, from
our point of view, not wholly bad.

In late June the unkept hedges are in the fulness of

their midsummer beauty. After sunset the fragrance
of the honeysuckle is almost too much : standing near
the blossom-laden hedge when there is no wind to
dissipate the odour, there is a heaviness in it which
makes it like some delicious honeyed liquor which we
are drinking in. The honeysuckle is indeed first among
the "melancholy flowers" that give out their fragrance by night.   In the daytime, when the smell is
faint, the pale sickly blossoms are hardly noticed even
where they are seen in masses and drape the hedges.
Of all the hedge-flowers, the rose alone is looked at, its
glory being so great as to make all other blooms seem
nothing but bleached or dead discoloured leaves in
comparison.

He would indeed be a vainly ambitious person who
should attempt to describe this queen of all wild
flowers, joyous or melancholy; but substituting flower
for fruit, and the delight of the eye for the pleasure of
taste, we may in speaking of it quote the words of a
famous old writer, used in praise of the strawberry.
He said that doubtless God Almighty could have made
a better berry if He had been so minded, but doubtless
God Almighty never did.

I esteem the rose not only for that beauty which sets
it highest among flowers, but also because it will not
suffer admiration when removed from its natural surroundings.   In this particular it resembles certain
brilliant sentient beings that languish and lose all their
charms in captivity.   Pluck your rose and bring it

indoors, and place it side by side with other blossoms—
yellow flag and blue periwinkle, and shining yellow
marsh-marigold, and poppy and cornflower—and it has
no lustre, and is no more to the soul than a flower made
out of wax or paper. Look at it here, in the brilliant
sunlight and the hot wind, waving to the wind on its
long thorny sprays all over the vast disordered hedges;
here in rosy masses, there starring the rough green
tangle with its rosy stars—a rose-coloured cloud on the
earth and Summer's bridal veil—and you will refuse to
believe (since it will be beyond your power to imagine)
that anywhere on the earth, in any hot or temperate climate, there exists a more divinely beautiful
sight.

If among the numberless cults that flourish in the
earth we could count a cult of the rose, to this spot the
votaries of the flower might well come each midsummer
to hold their festival. They would be youthful and
beautiful, their lips red, their eyes full of laughter; and
they would be arrayed in light silken garments of
delicate colour—green, rose, and white; and their arms
and necks and foreheads would shine with ornaments
of gold and precious stones. In their hands would be
musical instruments of many pretty shapes with which
they would sweetly accompany their clear voices as
they sat or stood beneath the old oak trees, and danced
in sun and shade, and when they moved in bright pro-
cession along the wide grass-grown roads, through
forest and farm-land.

In the summer of 1900 I found the lesser white-throat—the better whitethroat I should prefer to call it—in extraordinary abundance in the large unkept hedges east of the woods in the parishes of Fawley and Exbury. Hitherto I had always found this species everywhere thinly distributed; here it was as abundant as the reed-warblers along the dykes in the flat grass-lands on the Somerset coast, and like the reed-warblers in the reed and sedge-grown ditches and streams, each pair of whitethroats had its own part of the hedge; so that in walking in a lane when you left one singing behind you heard his next neighbour singing at a distance of fifteen or twenty yards farther on, and from end to end of the great hedge you had that continuous beautiful low warble at your side, and sometimes on both sides. The loud brief song of this whitethroat, which resembles the first part of a chaf-finch's song, is a pleasant sound and nothing more; the low warbling, which runs on without a break for forty or fifty seconds, or longer, is the beautiful song, and resembles the low continuous warble of the blackcap, but is more varied, and has one sound which is unique in the songs of British birds. This is a note repeated two or three times at intervals in the course of the song, of an excessive sharpness, unlike any other bird sound, but comparable to the silvery shrilling of the great green grasshopper—excessively sharp, yet musical. The bird emits this same silver shrill note when angry and when fighting, but it is then louder and not so musical,

E

and resembles the sharpest sounds made by bats and other small mammals when excited.

One day I sat down near a hedge, where an old half-dead oak stood among the thorns and brambles, and just by the oak a lesser whitethroat was moving about and singing. Out among the furze-bushes at some distance from the hedge a common whitethroat was singing, flitting and darting from bush to bush, rising at intervals into the air and dropping again into the furze; but by-and-by he rose to a greater height to pour out his mad confused strain in the air, then sloped away to the hedge and settled, still singing, on a dead branch of the oak. Up rose the lesser whitethroat and attacked it with extreme fury, rising to a height of two or three feet and dashing repeatedly at it, looking like a miniature kestrel or hobby; and every time it descended the other ducked his head and flattened himself on the branch, only to rise again, crest erect and throat puffed out, still pouring forth its defiant song. As long as this lasted the attacking bird emitted his piercing metallic anger-note, rapidly and continuously, like the clicking of steel machinery.

Alas! I fear I shall not again see the lesser white-throat as I saw him in that favoured year: in 1901 he came not, or came in small numbers; and it was the same in the spring of 1902. The spring was cold and backward in both years, and the bitter continuous east winds which prevailed in March and April probably proved fatal to large numbers of the more delicate migrants.

In this low, level country, sheltered by woods and hedgerows, we feel the tremendous power of the sun even before the last week in June. It is good to feel, to bathe in the heat all day long; but at noon one sometimes finds it too hot even on the open heath, and is forced to take shelter in the woods. It was always coolest on the high ground among the pines, where the trees are very tall and there is no underwood. In spring it was always pleasant to walk here on the thick carpet of fallen needles and old dead fern; now, in ·a very short time, the young bracken has sprung up as if by miracle to a nearly uniform height of about four feet. It spreads all around me for many acres an unbroken sea of brilliant green, out of which rise the tall red columns of the pines supporting the dark woodland roof.

Why is it, when in June the luxuriant young bracken first drops its fully-developed fronds, so that frond touches frond, many overlapping, forming a billowy expanse of vivid green, hiding, or all but hiding, the brown or red soil beneath—why is it the eyes rest with singular satisfaction on it? It is not only because of the colour, nor the beauty of contrast where the red floor of last year's beech leaves is seen through the fresh verdure, and of dark red-boled pines rising from the green sea of airy fronds. Colours and contrasts more beautiful may be seen, and the pleasure they give is different in kind.

Here standing amid the fern, where it had at last
formed that waving surface and was a little above my
knees, it seemed to me that the particular satisfaction
I experienced was due to the fine symmetrical leafing
of the surface, the minute subdivision of parts which
produced an effect similar to that of a mosaic floor.
When I consider other surfaces, on land or water, I
find the same gratification in all cases where it is
broken or marked out or fretted in minute, more or
less orderly subdivisions. The glass-like or oily surface
of water, where there are no reflections to bring other
feelings in, does not hold or attract but rather wearies
the sight; but it is no sooner touched to a thousand
minute crinkles by the wind, than it is looked at with
refreshment and pleasure. The bed of a clear stream,
with its pavement of minute variegated pebbles and
spots of light and shade, pleases in the same way.
The sight rests with some satisfaction even on a
stagnant pond covered with green duckweed; but the
satisfaction is less in this case on account of the
extreme minuteness of the parts and the too great
smoothness. The roads and open spaces in woods in
October and November are delightful to walk in when
they are like richly variegated floors composed of small
pieces, and like dark floors inlaid with red and gold
of beech and oak leaves. Numberless instances might
be given, and we see that the effect is produced even
in small objects, as, for instance, in scaly fishes and
in serpents. It is the minutely segmented texture of

BRACKEN IN JUNE

the serpent which, with the colour, gives it its wonderful richness. For the same reason a crocodile bag is more admired than one of cowhide, and a book in buckram looks better than one in cloth or even vellum.

The old Romans must have felt this instinctive pleasure of the eye very keenly when they took such great pains over their floors. I was strongly impressed with this fact at Silchester when looking at the old floors of rich and poor houses alike which have been uncovered during the last two or three years. They seem to have sought for the effect of mosaic even in the meaner habitations, and in passages and walks, and when tesseræ could not be had they broke up common tiles into small square fragments, and made their floors in that way. Even with so poor a material, and without any ornamentation, they did get the effect sought, and those ancient fragments of floors made of fragments of tiles, unburied after so many centuries, do actually more gratify the sight than the floors of polished oak or other expensive material which are seen in our mansions and palaces.

There is doubtless a physiological reason for this satisfaction to the eye, as indeed there is for so many of the pleasurable sensations we experience in seeing. We may say that the vision flies over a perfectly smooth plain surface, like a ball over a sheet of ice, and rests nowhere; but that in a mosaic floor the segmentation of the surface stays and rests the sight. To go no farther than that, which is but a part of the secret, the sheet of fern fronds, on account of this staying effect on the vision, increases what we see, so that a surface of a dozen square yards of fern seems more in extent than half an acre of smooth-shaven lawn, or the large featureless floor of a skating-rink or ball-room.

On going or wading through the belt of bracken under the tall firs—that billowy sea of fronds in the midst of which I have so long detained my patient reader—into the great oak wood beyond and below it, on each successive visit during the last days of June, the harshening of the bird voices became more marked. Only the wren and wood-wren and willow-wren uttered an occasional song, but the bigger birds made most of the sound. Families of young jays were then just out of the nest, crying with hunger, and filling the wood with their discordant screams when the parent birds came with food. A pair of kestrels, too, with a nestful of young on a tall fir incessantly uttered their shrill reiterated cries when I was near; and one pair of green woodpeckers, with young out of the breeding-hole but not yet able to fly, were half crazed with anxiety. Around me and on before me they flitted from tree to tree and clung to the bark, wings spread out and crest raised, their loud laugh changed to a piercing cry of anger that pained the sense.

They were now moved only by solicitude and anger: all other passion and music had gone out of the bird and into the insect world. The oak woods were now full of a loud continuous hum like that of a distant

AN ANGRY YAFFLE

threshing-machine; an unbroken deep sound composed of ten thousand thousand small individual sounds conjoined in one, but diffused and flowing like water over the surface, under the trees, and the rough bushy tangle. The incredible number and variety of blood-sucking flies makes this same low hot part of the Forest as nearly like a transcript of tropical nature in some damp, wooded district as may be found in England. But these Forest flies, even when they came in legions about me, were not able to spoil my pleasure. It was delightful to see so much life—to visit and sit down with them in their own domestic circle.

In other days, in a distant region, I have passed many a night out of doors in the presence of a cloud of mosquitoes; and when during restless sleep I have pulled the covering from my face, they had me at their mercy. For the smarts they inflicted on me then I have my reward, since the venom they injected into my veins has proved a lasting prophylactic. But to the poor cattle this place must be a very purgatory, a mazy wilderness swarming with minute hellish imps that mock their horns and giant strength, and cannot be shaken off. While sitting on the roots of a tree in the heart of the wood, I heard the heavy tramping and distressed bellowings of several beasts coming at a furious rate towards me, and presently half-a-dozen heifers and young bulls burst through the bushes; and catching sight of me at a distance of ten or twelve yards, they suddenly came to a dead stop, glaring at me with strange, mad, tortured eyes; then swerving aside, crashed away through the undergrowth in another direction.

In this wood I sought and found the stream well named the Dark Water; here, at all events, it is grown over with old ivied oaks, with brambles and briars that throw long branches from side to side, making the almost hidden current in the deep shade look black; but when the sunlight falls on it the water is the colour of old sherry from the red soil it flows over. No sooner had I sat down on the bank, where I had a little space of sunlit water to look upon, than the flies gathered thick about and on me, and I began to pay some attention to individuals among them. Those that came to suck blood, and settled at once in a business-like manner on my legs, were some hairy and some smooth, and of various colours—grey, black, steel-blue, and barred and ringed with bright tints; and with these distinguished guests came numberless others, small lean gnats mostly, without colour, and of no consideration.

I did not so much mind these as the others that simply buzzed round without an object — flies that have no beauty, no lancet to stab you with, and no distinction of any kind, yet will persist in forcing themselves on your attention. They buzz and buzz, and are loudest in your ear when you are most anxious to listen to some distant faint sound. If a blood-sucker hurts you, you can slap him to death, and there's an end of the matter; but slap at one of these idle, aimless, teasing flies as hard as you like, and he is gone like quick_ silver through your fingers. He is buzzing derisively in your ears: "Slap away as much as you like—it pleases you and doesn't hurt me." And then down again in the same place!

When the others—the serious flies on business bent —got too numerous, I began to slap my legs, killing one or two of the greediest at each slap, and to throw their small corpses on the sunlit current. These slain flies were not wasted, for very soon I had quite a number of little minnows close to my feet, eager to seize them as they fell. And, by-and-by, three fiddlers, or pond-skaters, "sagacious of their quarry from afar," came skating into sight on the space of bright water; and to these mysterious, uncanny - looking creatures —insect ghosts that walk on the water, but with very unghost-like appetites—I began tossing some of the flies; and each time a fiddler seized a floating fly he skated away into the shade with it to devour it in peace and quiet all alone by himself. For a fiddler with a fly is like a dog with a bone among other hungry dogs. When I had finished feeding my ghosts and little fishes, I got up and left the place, for the sun was travelling west and the greatest heat was over.

CHAPTER IV

DURING the last week in June we can look for the appearance of our most majestical insect, he is an evening flyer, and a little before sunset begins to show himself abroad. He is indeed a monarch among hexapods, with none to equal him save, perhaps, the great goblin moth ; and in shape and size and solidity he bears about the same relation to pretty bright flies as a horned rhinoceros does to volatile squirrels and monkeys and small barred and spotted felines. This is the stag-beetle — "stags and does" is the native name for the two sexes ; he is probably more abundant in this corner of Hampshire than in any other locality in England, and among the denizens of the Forest there are few more interesting. About four or five o'clock in the afternoon, the ponderous beetle wakes out of his long siesta, down among the roots and dead vegetable matter of a thorny brake or large hedge, and laboriously sets himself to work his way out. He is a slow, clumsy creature, a very bad climber; and small wonder, when we consider how he is impeded by his long branched horns when endeavouring to make his way upwards through a network of interlacing stems.

As you walk by the hedgeside a strange noise suddenly arrests your attention; it is the buzz of an insect, but loud enough to startle you; it might be mistaken for the reeling of a nightjar, but is perhaps more like the jarring hum of a fast driven motor-car. The reason of the noise is that the beetle has with great pains climbed up a certain height from the ground, and, in order to ascertain whether he has got far enough, he erects himself on his stand, lifts his wing-cases, shakes out his wings, and begins to agitate them violently, turning this way and that to make sure that he has a clear space. If he then attempts to fly—it is one of his common blunders—he instantly strikes against some branch or cluster of leaves, and is thrown down. The tumble does not hurt him in the least, but so greatly astonishes him that he remains motionless a good while; then recovering his senses, he begins to ascend again. At length, after a good many accidents and adventures by the way, he gets to a topmost twig, and, after some buzzing to get up steam, launches himself heavily on the air and goes away in grand style.

Hugh Miller, in his autobiography, tells of the dis-

covery he made of a curiously striking resemblance in
shape between our most elegantly made carriages and
the bodies of wasps, the resemblance being heightened
by a similarity of colouring seen in the lines and bands
of vivid yellows and reds on a polished black ground.
This likeness between insect and carriage does not
appear so striking at this day owing to a change in
the fashion towards a more sombre colour in the
vehicles; their funeral blacks, dark blues, and greens
being now seldom relieved with bright yellows and
reds. The stag-beetle, too, when he goes away with
heavy flight always gives one the idea of some kind
of machine or vehicle, not like the aerial phaeton of
the wasp or hornet, with its graceful lines and strongly-
contrasted colours, but an oblong, ponderous, armour-
plated car, furnished with a beak, and painted a deep
uniform brown.

Birds, especially the more aerial insectivorous kinds,
have the habit of flying at and teasing any odd or
grotesque-looking creature they may see on the wing—
as a bat, for instance. I have seen small birds dart at
a passing stag, but on coming near they turn tail and
fly from him, frightened perhaps at his formidable
appearance and loud noise.

Notwithstanding his lumbering, blundering ways,
when the stag is abroad in search of the doe, you
may see that he is endowed with a sense and faculty
so exquisite as to make it appear almost miraculous
in the sureness of its action. The void air, as he sweeps

droning through it, is peopled with subtle intelligences, which elude and mock and fly from him, and which he pursues until he finds out their secret. They mock him most, or, to drop the metaphor, he is most at fault, on a still sultry day when not a breath of air is stirring. At times he catches what, for want of better knowledge,

Stag-beetle.

we must call a scent, and in order to fix the direction it comes from he goes through a series of curious movements. You will see him rise above a thorny thicket, or a point where two hedges intersect at right angles, and remain suspended on his wings a few inches above the hedge-top for one or two minutes, loudly humming, and turning by a succession of jerks all round, pausing after each turn, until he has faced all points of the compass.

This failing, he darts away and circles widely round, then returning to the central point suspends himself as before. After spending several minutes in this manner, he once more resumes his wanderings. Several males are sometimes attracted to the same spot, but they pass and repass without noticing one another. You will see as many as three or four or half-a-dozen majestically moving up and down at a hedge side or in a narrow path in a hazel copse, each beetle turning when he gets to the end and marching back again, and altogether their measured, stately, and noisy movements are a fine spectacle.

A slight wind makes a great difference to him: even a current of air so faint as not to be felt on the face will reveal to him the exact distant spot in which the doe is lurking. The following incident will serve to show how perfect and almost infallible the sense and its correlated instinct are, and at the same time what a clumsy, blundering creature this beetle is.

Hearing a buzzing noise in a large unkept hedge, I went to the spot and found a stag trying to extricate himself from some soft fern fronds growing among the brambles in which he had got entangled. In the end he succeeded, and, finally gaining a point where there was nothing to obstruct his flight, he launched himself on the air and flew straight away to a distance of fifty yards; then he turned and commenced flying backwards and forwards, travelling forty or fifty yards one way and as many the other, until he made a discovery; and struck motionless in his career, he remained suspended for a moment or two, then flew swiftly and straight as a bullet back to the hedge from which he had so recently got away. He struck the hedge where it was broadest, at a distance of about twenty yards or more from the point where I had first found him, and running to the spot, I saw that he had actually alighted within four or five inches of a female concealed among the clustering leaves. On his approaching her she coyly moved from him, climbing up and down and along the branchlets, but for some time he continued very near her. So far he had followed on her track, or by the same branches and twigs over which she had passed, but on her getting a little farther away and doubling back, he attempted to reach her by a series of short cuts, over the little bridges formed by innumerable slender branches, and his short cuts in most cases brought him against some obstruction; or else there was a sudden bend in the branch, and he was taken farther away. When he had a chain of bridges or turnings, he seemed fated to take the wrong one, and in spite of all his desperate striving to get nearer, he only increased the distance between them. The level sun shone into the huge tangle of bramble, briar, and thorn, with its hundreds of interlacing branches and stringy stems, so that I was able to keep both beetles in sight; but after I had watched them for three-quarters of an hour, the sun departed, and I too left them. They were then nearly six feet apart; and seeing what a labyrinth they were in, I concluded that, strive how the enamoured creature might, they would never, from the stag-beetle point of view, be within measurable distance of one another.

Something in the appearance of the big beetle, both flying and when seen on the ground in his wrathful, challenging attitude, strikes the rustics of these parts as irresistibly comic. When its heavy flight brings it near the labourer in the fields, he knocks it down with his cap, then grins at the sight of the maltreated creature's amazement and indignation. However weary the ploughman may be when he plods his homeward way, he will not be too tired to indulge in this ancient practical joke. When the beetle's flight takes him by village or hamlet, the children, playing together in the road, occupied with some such simple pastime as rolling in the dust or making little miniature hills of loose sand, are suddenly thrown into a state of wild excitement, and, starting to their feet, they run whooping after the wanderer, throwing their caps to bring him down.

One evening at sunset, on coming to a forest gate through which I had to pass, I saw a stag-beetle standing in his usual statuesque, angry or threatening attitude in the middle of the road close to the gate. Doubtless some labourer who had arrived at the gate

F

earlier in the evening had struck it down for fun and
left it there. By-and-by, I thought, he will recover
from the shock to his dignity and make his way to
some elevated point, from which he will be able to start
afresh in his wanderings in search of a wife. But it was
not to be as I thought, for next morning, on going by
the same gate I found the remains of my beetle just
where I had last seen him—the legs, wing-cases, and
the big, broad head with horns attached. The poor
thing had remained motionless too long, and had been
found during the evening by a hedgehog and devoured,
all but the uneatable parts. On looking closely, I found
that the head was still alive; at a touch the antennæ—
those mysterious jointed rods, toothed like a comb at
their ends—began to wave up and down, and the horns
opened wide, like the jaws of an angry crab. On placing
a finger between them they nipped it as sharply as if
the creature had been whole and uninjured. Yet the
body had been long devoured and digested; and there
was only this fragment left, and, torn off with it, shall
we say? a fragment of intelligent life!

We always look on this divisibility of the life-principle
in some creatures with a peculiar repugnance; and, like
all phenomena that seem to contradict the regular course
of nature, it gives a shock to the mind. We do not
experience this feeling with regard to plant life, and to
the life of some of the lower animal organisms, because
we are more familiar with the sight in these cases. The
trouble to the mind is in the case of the higher life of

sentient and intelligent beings that have passions like our own. We see it even in some vertebrates, especially in serpents, which are most tenacious of life. Thus, there is a recorded case of a pit viper, the head of which was severed from the body by the person who found it. When the head was approached the jaws opened and closed with a vicious snap, and when the headless trunk was touched it instantly recoiled and struck at the touching object.

Such cases are apt to produce in some minds a sense as of something unfamiliar and uncanny behind nature that mocks us. But even those who are entirely free from any such animistic feeling are strangely disturbed at the spectacle, not only because it is opposed to the order of nature (as the mind apprehends it), but also because it contradicts the old fixed eternal idea we all have, that life is compounded of two things—the material body and the immaterial spirit, which leavens and, in a sense, recreates and shines in and through the clay it is mixed with; and that you cannot destroy the body without also destroying or driving out that myssterious, subtle principle. Life was thus anciently likened to a seal, which is two things in one—the wax and the impression on it. You cannot break the seal without also destroying the impression, any more than you can break a pitcher without spilling the liquor in it. In such cases as those of the beetle and the serpent, it would perhaps be better to liken life to a red, glowing ember, which may be broken into pieces, and each piece still burn and glow with its own portion of the original heat.

The survival after death of something commonly supposed to be dependent on vitality is another phenomenon which, like that of the divisibility of the life-principle, affects us disagreeably. The continued growth of the hair of dead men is an instance in point. It is, we know, an error, caused by the shrinking of the flesh; and as for the accounts of coffins being found full of hair when opened, they are inventions, though still believed in by some persons. Another instance, which is not a fable, is that of a serpent's skin. When properly and quickly dried after removal, it will retain its bright colours for an indefinite time—in some cases for many years. But at intervals the colours appear to fade, or become covered with a misty whiteness; and the cause, as one may see when the skin is rubbed or shaken, is that the outer scales are being shed. They come off separately, and are very much thinner than when the living serpent sheds his skin, and they grow thinner with successive sheddings until they are scarcely visible. But at each shedding the skin recovers its brightness. One in my possession continued shedding its scale films in this way for about ten years. I used it as a book-marker and often had it in my hands, but not until it ceased shedding its scale-coverings, and its original bright green colour turned to dull blackish-green, did I get rid of the feeling that it had some life in it.

But the most striking instance of the continuance or survival long after death of what has seemed an attribute or manifestation of life remains to be told.

One cloudy, very dark night at Boldre, I was going home across a heath with some girls from a farmhouse where we had been visiting, when one of my young companions cried out that she could see a spark of fire on the road before us. We then all saw it—a small, steady, green light—but on lighting a match and looking closely at the spot, nothing could we see except the loose soil in the road. When the match went out the spark of green fire was there still, and we searched again, turning the loose soil with our fingers until we discovered the dried and shrunken remains of a glow-worm of the previous year. It had been trodden into the sand, and the sand driven into it, until it was hard to make out any glow-worm shape or appearance in it. It was like a fragment of dry earth, and yet, so long as it was in the dark, the small, brilliant green light continued to shine from one end of it. Yet this dried old case must have been dead and blown about in the dust for at least seven or eight months.

On going up to London I carried it with me in a small box: there in a dark room it shone once more, but the light was now much fainter, and on the following evening there was no light. For some days I tried, by moistening it, by putting it out in the sun and wind, and in other ways, to bring back the light, but did not succeed; and, convinced at length that it would shine no more, I had the feeling that life had at last gone out of that dry, dusty fragment.

The little summer tragedies in Nature which we see or notice are very few—not one in a thousand of those that actually take place about us in a spot like this, teeming with midsummer life. A second one, which impressed me at the time, had for its scene a spot not more than eight minutes' walk from that forest gate where the stag-beetle, too long in cooling his wrath, had been overtaken by so curious a destiny. But before I relate this other tragedy, I must describe the place and some of the creatures I met there. It was a point where heath and wood meet, but do not mingle; where the marshy stream that drains the heath flows down into the wood, and the boggy ground sloping to the water is overgrown with a mixture of plants of different habits—lovers of a dry soil and of a wet—heather and furze, coarse and fine grasses, bracken and bog myrtle; and in the wettest spots there were patches and round masses of rust-red and orange-yellow and pale-grey lichen, and a few fragrant, shining, yellow stars of the bog asphodel, although its flowering season was nearly over. It was a perfect wilderness,

as wild a bit of desert as one could wish to be in,
where a man could spy all day upon its shy inhabitants, and no one would come and spy upon him.

Here, if anywhere, was my exulting thought when
I first beheld it, there should be adders for me.
There was a snakiness in the very look of the place,
and I could almost feel by anticipation the delightful
thrill in my nerves invariably experienced at the
sight of a serpent. And as I went very cautiously
along, wishing for the eyes of a dragon-fly so as to
be able to see all round me, a coil of black and
yellow caught my sight at a distance of a few yards
ahead, and was no sooner seen than gone. The spot
from which the shy creature had vanished was a
small, circular, natural platform on the edge of the
bank, surrounded with grass and herbage, and a little
dwarf, ragged furze; the platform was composed of
old, dead bracken and dry grass, and had a smooth,
flat surface, pressed down as if some creature used
it as a sleeping-place. It was, I saw, the favourite
sleeping- or basking-place of an adder, and by-and-by,
or in a few hours' time, I should be able to get a
good view of the creature. Later in the day, on
going back to the spot, I did find my adder on its
platform, and was able to get within three or four
yards, and watch it for some minutes before it slipped
gently down the bank and out of sight.

This adder was a very large (probably gravid) female,
very bright in the sunshine, the broad, zigzag band,

an inky black, on a straw-coloured ground. On my
third successful visit to the spot I was agreeably surprised to find that my adder had not been widowed
by some fatal accident, nor left by her wandering
mate to spend the summer alone; for now there were
two on the one platform, slumbering peacefully side
by side. The new-comer, the male, was a couple of
inches shorter and a good deal slimmer than his
mate, and differed in colour; the zigzag mark was
intensely black, as in the other, but the ground colour
was a beautiful copper red; he was, I think, the
handsomest red adder I have seen.

On my subsequent visits to the spot I found sometimes one and sometimes both; and I observed them
a good deal at different distances. One way was to
look at them from a distance of fifteen to twenty
yards through a binocular magnifying nine diameters,
which produced in me the fascinating illusion of
being in the presence of venomous serpents of a nobler
size than we have in this country. The glasses were
for pleasure only. When I watched them for profit
with my unaided eyes, I found it most convenient to
stand at a distance of three or four yards; but often
I moved cautiously up to the raised platform they
reposed on, until, by bending a little forward, I could
look directly down upon them.

When we first catch sight of an adder lying at rest
in the sun, it strikes us as being fast asleep, so motionless is it; but that it ever does really sleep with the

ADDERS

sun shining into its round, lidless, brilliant eyes is
hardly to be believed. The immobility which we
note at first does not continue long; watch the adder
lying peacefully in the sun, and you will see that at
intervals of a very few minutes, and sometimes as
often as once a minute, he quietly changes his position.
Now he draws his concentric coils a little closer, now
spreads them more abroad; by-and-by the whole
body is extended to a sinuous band, then disposed in
the form of a letter S, or a simple horseshoe figure,
and sometimes the head rests on the body and
sometimes on the ground. The gentle, languid movements of the creature changing his position at intervals are like those of a person in a reclining hot
bath, who occasionally moves his body and limbs to
renew and get the full benefit of the luxurious sensation.

That the two adders could see me when I stood
over them, or at a distance of three or four yards, or
even more, is likely; but it is certain that they did
not regard me as a living thing, or anything to be
disturbed at, but saw me only as a perfectly motionless
object which had grown imperceptibly on their vision,
and was no more than a bush, or stump, or tree.
Nevertheless, I became convinced that always after
standing for a time near them my presence produced
a disturbing effect. It is, perhaps, the case that we
are not all contained within our visible bodies, but
have our own atmosphere about us — something of

us which is outside of us, and may affect other creatures. More than that, there may be a subtle current
which goes out and directly affects any creature (or
person), which we regard for any length of time with
concentrated attention. This is one of the things
about which we know nothing, or, at all events, learn
nothing from our masters, and most scientists would
say that it is a mere fancy; but in this instance it
was plain to see that always after a time *something*
began to produce a disturbing effect on the adders.
This would first show itself in a slight restlessness,
a movement of the body as if it had been breathed
upon, increasing until they would be ill at ease all
the time, and at length they would slip quietly away
to hide under the bank.

The following incident will show that they were
not disturbed at seeing me standing near, assuming
that they could or did see me. On one of my visits
I took some pieces of scarlet ribbon to find out by
an experiment if there was any truth in the old belief
that the sight of scarlet will excite this serpent to
anger. I approached them in the usual cautious way,
until I was able, bending forward, to look down upon
them reposing unalarmed on their bed of dry fern;
then, gradually putting one hand out until it was
over them, I dropped from it first one then another
piece of silk so that they fell gently upon the edge
of the platform. The adders must have seen these
bright objects so close to them, yet they did not sud-

denly draw back their heads, nor exsert their tongues,
nor make the least movement, but it was as if a
dry, light, dead leaf, or a ball of thistledown, had
floated down and settled near them, and they had
not heeded it.

In the same way they probably saw me, and it
was as if they had seen me not, since they did not
heed my motionless figure; but that they always
felt my presence after a time I felt convinced, for
not only when I stood close to and looked down
upon them, but also at a distance of four to eight
yards, after gazing fixedly at them for some minutes,
the change, the tremor, would appear, and in a little
while they would steal away.

Enough has been said to show how much I liked
the company of these adders, even when I knew that
my presence disturbed their placid lives in some in-
definable way. They were indeed more to me than all
the other adders, numbering about a score, which I
had found at their favourite basking places in the
neighbourhood. For they were often to be found in
that fragrant, sequestered spot where their home was,
and they were two together, of different types, both
beautiful, and by observing them day by day I increased my knowledge of their kind. We do not
know very much about "the life and conversation" of
adders, having been too much occupied in "bruising"
their shining beautiful bodies beneath our ironshod
heels, and with sticks and stones, to attend to such

matters. So absorbed was I in contemplating or else thinking about them at that spot that I was curiously indifferent to the other creatures—little lizards, and butterflies, and many young birds brought by their parents to the willows and alders that shaded the stream. All day the birds dozed on their gently-swaying perches, chirping at intervals to be fed; and near by a tree-pipit had his stand, and sang and sang when most songsters were silent, but I paid no attention even to his sweet strains. Two or three hundred yards away, up the stream on a boggy spot, a pair of peewits had their breeding-place. They were always there, and invariably on my appearance they rose up and came to me, and, winnowing the air over my head, screamed their loudest. But I took no notice, and was not annoyed, knowing that their most piercing cries would have no effect on the adders, since their deaf ears heard nothing, and their brilliant eyes saw next to nothing of all that was going on about them. After vexing their hearts in vain for a few minutes the peewits would go back to their own ground, then peace would reign once more.

One day I was surprised and a little vexed to find that the peewits had left their own ground to come and establish themselves on the bog within forty yards of the spot where I was accustomed to take my stand when observing the adders. Their anxiety at my presence had now become so intensified that it was painful to witness. I concluded that they had led their nearly grown-up young to that spot, and sincerely hoped that they would be gone on the morrow. But they remained there five days; and as their solicitude and frantic efforts to drive me away were renewed on each visit, they were a source of considerable annoyance. On the fourth day I accidentally discovered their secret. If I had not been so taken up with the adders, I might have guessed it. Going over the ground I came upon a dead full-grown young peewit, raised a few inches above the earth by the heather it rested on, its head dropped forward, its motionless wings partly open.

Usually at the moment of death a bird beats violently with its wings, and after death the wings remain half open. This was how the peewit had died, the wings half folded. Picking it up, I saw that it had been dead several days, though the carrion beetles had not attacked it, owing to its being several inches above the ground. It had, in fact, no doubt been already dead when I first found the old peewits settled at that spot; yet during those four hot, long summer days they had been in a state of the most intense anxiety for the safety of these dead remains! This is to my mind not only a very pathetic spectacle, but one of the strangest facts in animal life. The reader may say that it is not at all strange, since it is very common. It is most strange to me because it is very common, since if it were rare we could say that it

was due to individual aberration, or resulted through
the bluntness of some sense or instinct.   What is
wonderful and almost incredible is that the higher
vertebrates have no instinct to guide them in such a
case as I have described, and no inherited knowledge
of death.   To make of Nature a person, we may see
that in spite of her providential care for all her children, and wise ordering of their lives down to the
minutest detail, she has yet failed in this one thing.
Her only provision is that the dead shall be speedily
devoured; but they are not thus removed in numberless instances; a very familiar one is the sight of
living and dead young birds, the dead often in a
state of decay, lying together in one nest: and here
we cannot but see that the dead become a burden
and a danger to the living.   Birds and mammals are
alike in this.   They will call, and wait for, and bring
food to, and try to rouse the dead young or mate;
day and night they will keep guard over it and waste
themselves in fighting to save it from their enemies.
Yet we can readily believe that an instinct fitted to
save an animal from all this vain excitement, and
labour, and danger, would be of infinite advantage to
the species that possessed it.

In some social hymenopterous insects we see that
the dead are removed; it would be impossible for ants
to exist in communities numbering many thousands
and tens of thousands of members crowded in a small
space without such a provision.   The dead ant is picked

up by the first worker that happens to come that way and discovers it, and carried out and thrown away. Probably some chemical change which takes place in the organism on the cessation of life and makes it offensive to the living has given rise to this healthy instinct. The dead ant is not indeed seen as a dead fellow-being, but as so much rubbish, or "matter in the wrong place," and is accordingly removed. We can confidently say that this is not a knowledge of death, from what has been observed of the behaviour of ants on the death of some highly regarded individual in the nest—a queen, for instance. On this point I will quote a passage from the Rev. William Gould's *Account of English Ants*, dated 1747. His small book may be regarded as a classic, at all events by naturalists; albeit the editors of our *Dictionary of National Biography* have not thought proper to give him a place in that work, in which so many obscurities, especially of the nineteenth century, have had their little lives recorded.

It may be remarked in passing that the passage to be quoted is a very good sample of the style of our oldest entomologist, the first man in England to observe the habits of insects. His small volume dates many years before the *Natural History of Selborne*, and his style, it will be seen, is very different from that of Gilbert White. We know from Lord Avebury's valuable book on the habits of ants that Gould was not mistaken in these remarkable observations.

"In whatever Apartment a Queen Ant condescends to be present, she commands Obedience and Respect. An universal Gladness spreads itself through the whole Cell, which is expressed by particular Acts of Joy and Exultation. They have a peculiar Way of skipping, leaping, and standing upon their Hind Legs, and prancing with the others. These Frolicks they make use of, both to congratulate each other when they meet, and to show their regard for the Queen. . . . However Fantastick this Description may appear, it may easily be proved by an obvious Experiment. If you place a Queen Ant with her Retinue under a Glass, you will in a few Moments be convinced of the Honour they pay, and Esteem they Entertain for her. There cannot be a more remarkable historie than what happened to a Black Queen the beginning of last Spring. I had placed her with a bare Retinue in a sliding Box, in the cover of which was an opening sufficient for the workers to pass to and fro, but so narrow as to confine the Queen. A Corps was constantly in waiting and surrounded her, whilst others went out in search of Provisions. By some Misfortune she died: the Ants, as if not apprised of her Death, continued their Obedience. They even removed her from one part of the Box to another, and treated her with the same Court and Formality as if she had been alive. This lasted two Months, at the end of which, the Cover being open, they forsook the Box, and carried her off."

Two days after I found the dead peewit the parent birds disappeared; and a little later I paid my last visit to the adders, and left them with the greatest reluctance, for they had not told me a hundredth part of their unwritten history.

CHAPTER V

THE nightingale ceases singing about June 18 or 20. A bird here and there may sing later; I occasionally hear one as late as the first days of July. And because the nightingale is not so numerous as the other singers, and his song attracts more attention, we get the idea that his musical period is soonest over. Yet several other species come to the end of their vocal season quite as early, or but little later. If it be an extremely abundant species, as in the case of the willow-wren, we will hear a score or fifty sing for every nightingale. Blackcap and garden warbler, whitethroat and lesser whitethroat, are nearly silent, too, at the beginning of July; and altogether it seems to be the rule that the species oftenest heard after June are the most abundant.

The woodland silence increases during July and August, not only because the singing season is ended, but also because the birds are leaving the woods: that darkness and closeness which oppress us when we walk in the deep shade is not congenial to them; besides, food is less plentiful than in the open places, where the sun shines and the wind blows.

Woods, again, vary greatly in character and the degree of attractiveness they have for birds: the copse and spinney keep a part of their population through the hottest months; and coming to large woods the oak is never oppressive like the beech and other deciduous trees. It spreads its branches wide, and has wide spaces which let in the light and air; grass and undergrowth flourish beneath it, and, better than all, it abounds in bird food on its foliage above all trees.

My favourite woods were almost entirely of oak with a holly undergrowth, and at some points oaks were mixed with firs. They were never gloomy nor so silent as most woods; but in July, as a rule, one had to look for the birds, since they were no longer distributed through the wood as in the spring and early summer, but were congregated at certain points.

Most persons are familiar with those companies of small birds which form in woods in winter, composed of tits of all species, with siskins, goldcrests, and sometimes other kinds. The July gatherings are larger, include more species, and do not travel incessantly like the winter companies. They are composed of families

—parent birds and their young, lately out of the nest, brought to the oaks to be fed on caterpillars. It may be that their food is more abundant at certain points, but it is also probable that their social disposition causes them to congregate. Walking in the silent woods you begin to hear them at a considerable distance ahead—a great variety of sounds, mostly of that shrill, sharp, penetrative character which is common to many young passerine birds when calling to be fed. The birds will sometimes be found distributed over an acre of ground, a family or two occupying every large oak tree—tits, finches, warblers, the tree-creeper, nuthatch, and the jay. What, one asks, is the jay doing in such company? He is feeding at the same table, and certainly not on them. All, jays included, are occupied with the same business, minutely examining each cluster of leaves, picking off every green caterpillar, and extracting the chrysalids from every rolled-up leaf. The airy little leaf warblers and the tits do this very deftly; the heavier birds are obliged to advance with caution along the twig until by stretching the neck they can reach their prey lurking in the green cluster, and thrust their beaks into each little green web-fastened cylinder. But all are doing the same thing in pretty much the same way. While the old birds are gathering food, the young, sitting in branches close by, are incessantly clamouring to be fed, their various calls making a tempest of shrill and querulous sounds in the wood. And the shrillest of all are the long-tailed tits; these will not sit still and wait like the others, but all, a dozen or fifteen to a brood, hurry after their busy parents, all the time sending out those needles of sound in showers. Of hard-billed birds the chaffinch, as usual, was the most numerous, but there were, to my surprise, many yellowhammers, all these, like the rest, with their newly brought out young. The presence of the hawfinch was another surprise; and here I noticed that the hunger call of the young hawfinch is the loudest of all—a measured, powerful, metallic chirp, heard high above the shrill hubbub.

Watching one of these busy companies of small birds at work one is amazed at the thought of the abundance of larval insect life in these oak woods. The caterpillars must be devoured in tens of thousands every day for some weeks, yet when the time comes one is amazed again at the numbers that have survived to know a winged life. On July evenings with the low sun shining on the green oaks at this place I have seen the trees covered as with a pale silvery mist—a mist composed of myriads of small white and pale-grey moths fluttering about the oak foliage. Yet it is probable that all the birds eat is but a small fraction of the entire number destroyed. The rapacious insects are in myriads too, and are most of them at war with the soft-bodied caterpillars. The earth under the bed of dead leaves is full of them, and the surface is hunted over all day by the wood or horse ants—Formica rufa. One day, standing still to watch a number of these ants

moving about in all directions over the ground, I saw a
green geometer caterpillar fall from an oak leaf above
to the earth, and no sooner had it dropped than an ant
saw and attacked it, seizing it at one end of its body
with his jaws. The caterpillar threw itself into a horseshoe form and then violently jerking its body round
flung the ant away to a distance of a couple of inches.
But the attack was renewed, and three times the ant
was thrown violently off; then another ant came, and
he, too, was twice thrown off; then a third ant joined in
the fight, and when all three had fastened their jaws on
their victim the struggle ceased, and the caterpillar was
dragged away. That is the fate of most caterpillars
that come to the ground. But the ants ascend the
trees; you see them going up and coming down in
thousands, and you find on examination that they dis-
tribute themselves over the whole tree, even to the
highest and farthest terminal twigs. And their numbers are incalculable—here in the Forest, at all events.
Not only are their communities large, numbering
hundreds of thousands in a nest, but their nests here
are in hundreds, and it is not uncommon to find them
in groups, three or four up to eight or ten, all within a
distance of a few yards of one another.

I had thought to write more, a whole chapter in fact,
on this fascinating and puzzling insect—our "noble
ant," as our old ant lover Gould called it; but I have
had to throw out that and much besides in order to
keep this book within reasonable dimensions.

There is another noise of birds in all woods and copses in the silent season which is familiar to every one—the sudden excited cries they utter at the sight of some prowling animal—fox, cat, or stoat. Even in the darkest, stillest woods these little tempests of noise occasionally break out, for no sooner does one bird utter the alarm cry than all within hearing hasten to the spot to increase the tumult. These tempests are of two kinds—the greater and lesser; in the first jays, blackbirds, and missel-thrushes take part, the magpie, too, if he is in the wood, and almost invariably the outcry is caused by the appearance of one of the animals just named. In the smaller outbreaks, which are far more frequent, none of these birds take any part, not even the excitable blackbird, in spite of his readiness to make a noise on the least provocation. Only the smaller birds are concerned here, from the chaffinch down; and the weasel is, I believe, almost always the exciting cause. If it be as I think, a curious thing is that birds like the chaffinch and the tits, which have their nests placed out of its reach, should be so overcome at the sight of this minute creature which hunts on the ground, and which blackbirds and jays refuse to notice in spite of the outrageous din of the finches. The chaffinch is invariably first and loudest in these outbreaks; a dozen or twenty times a day, even in July and August, you will hear his loud passionate pink-pink calling on all of his kind to join him, and by-and-by, if you can succeed in getting to the spot, you will hear other species joining in—the girding of ox-eye and blue-tit, the angry, percussive note of the wren, the low wailing of the robin, and the still sadder dunnock, and the small plaintive cries of the tree warblers.

What an idle demonstration, what a fuss about nothing it seems! The minute weasel is on the track of a vole or a wood-mouse and cannot harm the birds. Yes, he can take the nestlings from the robin's and willow-wren's nests, and from other nests built on the ground, but what has the chaffinch to do with it all? Can it be that there is some fatal weakness in birds, in spite of their wings, in this bird especially, such as exists in voles, and mice, and rabbits, and in frogs and lizards, which brings them down to destruction, and of which they are in some way conscious? Some months ago there was a correspondence in the *Field* which touched upon this very subject. One gentleman wrote that he had found three freshly-killed adult cock chaffinches in a weasel's nest, and he asked in consequence how this small creature that hunts on the ground could be so successful in capturing so alert and vigorous a bird as this finch?

For a long time before this correspondence appeared I had been trying to find out the secret of the matter, but the weasel has keen senses, and it is hard to see and follow his movements in a copse without alarming him. One day, over a year ago, near Boldre, I was fortunate enough to hear a commotion of the lesser kind at a spot where I could steal upon without alarming the little beast. There was an oak tree, with some scanty thorn-bushes growing beside the trunk, and stealing quietly to the spot I peeped through the screening thorns, and saw a weasel lying coiled round, snakewise, at the roots of the oak in a bed of dead leaves. He was grinning and chattering at the birds, his whole body quivering with excitement. Close to him on the twigs above the birds were perched, and fluttering from twig to twig—chaffinches, wrens, robins, dunnocks, ox-eyes, and two or three willow-wrens and chiffchaffs. The chaffinches were the most excited, and were nearest to him. Suddenly, after a few moments, the weasel began wriggling and spinning round with such velocity that his shape became indistinguishable, and he appeared as a small round red object violently agitated, his rapid motions stirring up the dead leaves so that they fluttered about him. Then he was still again, but chattering and quivering, then again the violent motion, and each time he made this extraordinary movement the excitement and cries of the birds increased and they fluttered closer down on the twigs. Unluckily, just when I was on the point of actually witnessing the end of this strange little drama—a chaffinch, I am sure, would have been the victim—the little flat-headed wretch became aware of my presence, not five yards from him, and springing up he scuttled into hiding.

If, as I think, certain species of birds are so thrown off their mental balance by the sight of this enemy as to come in their frenzy down to be taken by him, it is clear that he fascinates—to use the convenient old word—in two different ways, or that his furred and feathered victims are differently affected. In the case of the rabbits and of the small rodents, we see that they recognise the dangerous character of their pursuer and try their best to escape from him, but that they cannot attain their normal speed—they cannot run as they do from a man, or dog, or other enemy, or as they run ordinarily when chasing one another. Yet it is plain to any one who has watched a rabbit followed by a stoat that they strain every nerve to escape, and, conscious of their weakness, are on the brink of.despair and ready to collapse. The rabbit's appearance when he is being followed, even when his foe is at a distance behind, his trembling frame, little hopping movements, and agonising cries, which may be heard distinctly three or four hundred yards away, remind us of our own state in a bad dream, when some terrible enemy, or some nameless horror, is coming swiftly upon us; when we must put forth our utmost speed to escape instant destruction, yet have a leaden weight on our limbs that prevents us from moving.

I have often watched rabbits hunted by stoats, and recently, at Beaulieu, I watched a vole hunted by a weasel, and it was simply the stoat and rabbit hunt in little.

It is a typical case, and I will describe just what I saw, and saw very well. I was on the hard, white road between Beaulieu village and Hilltop, when the

WEASEL FASCINATING SMALL BIRDS

little animal—a common field vole—came out from the hedge and ran along the road, and knowing from his appearance that he was being pursued, I stood still to see the result. He had a very odd look: instead of a smooth-haired little mouse-like creature running smoothly and swiftly over the bare ground, he was all hunched up, his hair standing on end like bristles, and he moved in a series of heavy painful hops. Before he had gone half-a-dozen yards, the weasel appeared at the point where the vole had come out, following by scent, his nose close to the ground; but on coming into the open road he lifted his head and caught sight of the straining vole, and at once dashed at and overtook him. A grip, a little futile squeal, and all was over, and the weasel disappeared into the hedge. But his mate had crossed the road a few moments before—I had seen her run by me—and he wanted to follow her, and so presently he emerged again with the vole in his mouth, and plucking up courage ran across close to me. I stood motionless until he was near my feet, then suddenly stamped on the hard road, and this so startled him that he dropped his prey and scuttled into cover. Very soon he came out again, and, seeing me so still, made a dash to recover his vole, when I stamped again, and he lost it again and fled; but only to return for another try, until he had made at least a dozen attempts. Then he gave it up, and peering at me in a bird-like way from the roadside grass began uttering a series of low, sorrowful sounds, so low indeed that if I had been more than six yards from him they would have been inaudible—low, and soft, and musical, and very sad, until he quite melted my heart, and I turned away, leaving him to his vole, feeling as much ashamed of myself as if I had teased a pretty bright-eyed little child by keeping his cake or apple until I had made him cry.

With regard to these fatal weaknesses in birds, mammals, and reptiles, which we see are confined to certain species, they always strike us as out of the order of nature, or as abnormal, if the word may be used in such a connection. Perhaps it can be properly used. I remember that Herodotus, in his *History of Egypt*, relates that when a fire broke out in any city in that country, the people did not concern themselves about extinguishing it; their whole anxiety was to prevent the cats from rushing into the flames and destroying themselves. To this end the people would occupy all the approaches to the burning building, forming a cordon, as it were, to keep the cats back; but in spite of all they could do, some of them would get through, and rush into the flames and die. The omniscient learned person may tell me that Herodotus is the Father of Lies, if he likes, and is anxious to say something witty and original; but I believe this story of the cats, since not Herodotus, nor any Egyptian who was his informant, would or could have invented such a tale. Believing it, I can only explain it on the assumption that this Egyptian race of cats had become subject to a fatal weakness, a hypnotic effect caused by the sight of a great blaze. In like manner, if our chaffinch gets too much excited and finally comes down to be destroyed by a weasel, when he catches sight of that small red animal, or sees him going through that strange antic performance which I witnessed, it does not follow that the weakness or abnormality is universal in the species. It may be only in a race.

Again, with regard to rabbits: when hunted by a stoat they endeavour to fly, but cannot, and are destroyed owing to that strange—one might almost say unnatural — weakness; but I can believe that if a colony of British rabbits were to inhabit, for a good many generations, some distant country where there are no stoats, this weakness would be outgrown. It is probable that, even in this stoat-infested country, not all individuals are subject to such a failing, and that in those which have it, it differs in degree. If it is a weakness, a something inimical, then it is reasonable to believe that nature works to eliminate it, whether by natural selection or some other means.

The main point is the origin of this flaw in certain races, and perhaps species. How comes it that certain animals should, in certain circumstances, act in a definite way, as by instinct, to the detriment of their own and the advantage of some other species—in this case that of a direct and well-known enemy? It is

a mystery, one which, so far as I know, has not yet
been looked into. A small ray of light may be thrown
on the matter, if we consider the fact of those strange
weaknesses and mental abnormalities in our own
species, which are supposed to have their origin in
violent emotional and other peculiar mental states in
one of our parents. The fathers have eaten sour
grapes, and their children's teeth are set on edge, is
one of the old proverbs quoted by Ezekiel. I know
of one unfortunate person who, if he but sees a lemon
squeezed, or a child biting an unripe-looking fruit,
has his teeth so effectually set on edge, that he cannot
put food into his mouth for some time after. Here is a
farmer, a big, strong, healthy man, who himself works
on his farm like any labourer, who, if he but catches
sight of any ophidian—adder, or harmless grass snake,
or poor, innocent blindworm—instantly lets fall the
implements from his hands, and stands trembling,
white as a ghost, for some time; then, finally, he goes
back to the house, slowly and totteringly, like some
very aged, feeble invalid, and dropping on to a bed, he
lies nerveless for the rest of that day. Night and sleep
restore him to his normal state.

I give this one of scores of similar cases which I
have found. Such things are indeed very common.
But how does the fact of pre-natal suggestion help us
to get at the true meaning of such a phenomenon as
fascination? It does not help us if we consider it by
itself. It is a fact that "freaks" of this kind, mental

and physical, are transmissible, but that helps us little
—the abnormal individual has the whole normal race
against him. Thus, in reference to the cat story in
Herodotus, here in a Hampshire village, a mile or two
from where I am writing this chapter, a cottage took
fire one evening, and when the villagers were gathered
on the spot watching the progress of the fire, some
pigs—a sow with her young ones—appeared on the
scene and dashed into the flames. The people rushed
to the rescue, and with some difficulty pulled the pigs
out; and finally hurdles had to be brought and placed
in the way of the sow to prevent her getting back, so
anxious was she to treat the villagers to roast pig.

This is a case of the hypnotic effect of fire on
animals, and perhaps many similar cases would be
found if looked for. We know that most animals
are strangely attracted by fire at night, but they fear
it too, and keep at a proper distance. It draws and
disturbs but does not upset their mental balance. But
how it came about that a whole race of cats in ancient
Egypt were thrown off their balance and were always
ready to rush into destruction like the Hampshire pigs,
is a mystery.

To return to fascination. Let us (to personify) remember that Nature in her endeavours to safeguard
all and every one of her creatures has given them the
passion of fear in various degrees, according to their
several needs, and in the greatest degree to her persecuted weaklings; and that this emotion, to be efficient,

must be brought to the extreme limit, beyond which it becomes debilitating and is a positive danger, even to betraying to destruction the life it was designed to save. Let us consider this fact in connection with that of pre-natal suggestion — of weak species frequently excited to an extremity of fear at the sight, familiar to them, of some deadly enemy, and the possible effect of that constantly recurring violent disturbance and image of terror on the young that are to be.

The guess may go for what it is worth. We know that the susceptibility of certain animals—the vole and the frog, let us say—to fascination, is like nothing else in animal life, since it is a great disadvantage to the species, a veritable weakness, which might even be called a disease; and that it must therefore have its cause in too great a strain on the system somewhere; and we know, too, that it is inheritable. But the facts are too few, since no one has yet taken the pains to collect data on the matter. There is a good deal of material lying about in print; and I am astonished at many things I hear from intelligent keepers, and other persons who see a good deal of wild life, bearing on this subject. But I do not now propose to follow it any further.

I went into the oak wood one morning, and, finding it unusually still, betook myself to a spot where I had often found the birds gathered. It was a favourite

place, where there was running water and very large
trees standing wide apart, with a lawn-like green turf
beneath them. This green space was about half an
acre in extent, and was surrounded by a thicker wood
of oak and holly, with an undergrowth of brambles.
Here I found a dead squirrel lying on the turf under
one of the biggest oaks, looking exceedingly conspicuous
with the bright morning sun shining on him.

A poor bag! the reader may say, but it was the day
of small things at the end of July, and this dead
creature gave me something to think about. How in
the name of wonder came it to be dead at that peaceful place, where no gun was fired! I could not believe
that he had died, for never had I seen a finer, glossier-
coated, better-nourished-looking squirrel. "Whiter
than pearls are his teeth," were Christ's words in the
legend when His followers looked with disgust and
abhorrence at a dead dog lying in the public way.
This dead animal had more than pearly teeth to
admire; he was actually beautiful to the sight, lying
graceful in death on the moist green sward in his rich
chestnut reds and flower-like whiteness. The wild,
bright-eyed, alert little creature—it seemed a strange
and unheard of thing that he, of all the woodland
people, should be lying there, motionless, not stiffened
yet and scarcely cold.

A keeper in Hampshire told me that he once saw a
squirrel accidentally kill itself in a curious way. The
keeper was walking on a hard road, and noticed the

H

squirrel high up in the topmost branches of the trees overhead, bounding along from branch to branch before him, and by-and-by, failing to grasp the branch it had aimed at, it fell fifty or sixty feet to the earth, and was stone dead when he picked it up from the road. But such accidents must be exceedingly rare in the squirrel's life.

Looking closely at my dead squirrel to make sure that he had no external hurt, I was surprised to find its fur peopled with lively big black fleas, running about as if very much upset at the death of their host. These fleas were to my eyes just like pulex irritans — our own flea; but it is doubtful that it was the same, as we know that a great many animals have their own species to tease them. Now, I have noticed that some very small animals have very small fleas; and that, one would imagine, is as it should be, since fleas are small to begin with, because they cannot afford to be large, and the flea that would be safe on a dog would be an unsuitable parasite for so small a creature as a mouse. The common shrew is an example. It has often happened that when in an early morning walk I have found one lying dead on the path or road and have touched it, out instantly a number of fleas have jumped. And on touching it again, there may be a second and a third shower. These fleas, parasitical on so minute a mammal, are themselves minute—pretty sherry-coloured little creatures, not half so big as the dog's flea. It appears

to be a habit of some wild fleas, when the animal
they live on dies and grows cold, to place themselves
on the surface of the fur and to hop well away when
shaken. But we do not yet know very much about
their lives. Huxley once said that we were in danger
of being buried under our accumulated monographs.
There is, one is sorry to find, no monograph on the
fleas; a strange omission, when we consider that

A dead Shrew.

we have, as the life-work of an industrious German,
a big handsome quarto, abundantly illustrated, on
the more degraded and less interesting *Pedicularia*.

The multitude of fleas, big and black, on my dead
squirrel, seemed a ten times bigger puzzle than the
one of the squirrel's death. For how had they got
there? They were not hatched and brought up on
the squirrel: they passed their life as larvæ on the
ground, among the dead leaves, probably feeding on

decayed organic matter. How did so many of them succeed in getting hold of so very sprightly and·irritable a creature, who lives mostly high up in the trees, and does not lie about on the ground ? Can it be that fleas—those proper to the squirrel—swarm on the ground in the woods, and that without feeding on mammalian blood they are able to propagate and keep up their numbers ? These questions have yet to be answered.

It struck me at last that these sprightly parasites might have been the cause of the squirrel's coming to grief; that, driven to desperation by their persecutions, he had cast himself down from some topmost branch, and so put an end to the worry with his life.

Squirrels abound in these woods, and but for parasites and their own evil tempers they might be happy all the time. But they are explosive and tyrannical to an almost insane degree ; and this may be an effect of the deleterious substances they are fond of eating. They will feast on scarlet and orange agarics—lovely things to look at, but deadly to creatures that are not immune. A prettier spectacle than two squirrels fighting is not to be seen among the oaks. So swift are they, so amazingly quick in their doublings, in feints, attack, flight, and chase ; moving not as though running on trees and ground, but as if flying and gliding ; and so rarely do they come within touching distance of one another, that the delighted looker-on might easily suppose that it is all in fun. In their most truculent moods, in their fiercest fights, they cannot cease to be graceful in all their motions.

A common action of squirrels, when excited, of throwing things down, has been oddly misinterpreted by some observers who have written about it. Here I have often watched a squirrel, madly excited at my presence when I have stopped to watch him, dancing about and whisking his tail, scolding in a variety of tones, and emitting that curious sound which reminds one of the chattering cry of fieldfares when alarmed; and finally tearing off the loose bark with his little hands and teeth, and biting, too, at twigs and leaves so as to cause them to fall in showers. The little pot boils over in that way, and that's all there is to be said about it.

Walking among the oaks one day in early winter when the trees were nearly leafless, I noticed a squirrel sitting very quietly on a branch; and though he did not get excited, he began to move away before me, stopping at intervals and sitting still to watch me for a few moments. He was a trifle suspicious, and nothing more. In this way he went on for some distance, and by-and-by came to a long horizontal branch thickly clothed with long lichen on its upper sides, and instantly his demeanour changed. He was all excitement, and bounding along the branch he eagerly began to look for something, sniffing and scratching with his paws, and presently he pulled

out a nut which had been concealed in a crevice
under the lichen, and sitting up, he began cracking
and eating it, taking no further notice of me. The
sudden change in him, the hurried search for something, and the result, seemed to throw some light
on the question of the animal's memory with reference to his habit of hiding food. It is one common
to a great number of rodents, and to many of the
higher mammals—Canidæ and Felidæ, and to many
birds, including most, if not all, the Corvidæ.

When the food is hidden away here, there, and
everywhere, we know from observation that in innumerable instances it is never found, and probably
never looked for again; and of the squirrel we are
accustomed to say that he no sooner hides a nut
than he forgets all about it. Doubtless he does, and
yet something may bring it back to his mind. In
this matter I think there is a considerable difference
between the higher mammals, cats and dogs, for instance, and the rodents; I think the dog has a better
or more highly developed memory. Thus, I have
seen a dog looking enviously at another who had
got a bone, and after gazing at him with watering
mouth for some time, suddenly turn round and go
off at a great pace to a distant part of the ground,
and there begin digging, and presently pull out a
bone of his own, which he had no doubt forgotten
all about until he was feelingly reminded of it. I
doubt if a squirrel would ever rise to this height;

but on coming by chance to a spot with very marked features, where he had once hidden a nut, then I think the sight of the place might bring back the old impression.

I have often remarked when riding a nervous horse, that he will invariably become alarmed, and sometimes start at nothing, on arriving at some spot where something had once occurred to frighten him. The sight of the spot brings up the image of the object or sound that startled him; or, to adopt a later interpretation of memory, the past event is reconstructed in his mind. Again, I have noticed with dogs, when one is brought to a spot where on a former occasion he has battled with or captured some animal, or where he has met with some exciting adventure, he shows by a sudden change in his manner, in eyes and twitching nose, that it has all come back to him, and he appears as if looking for its instant repetition.

We see that we possess this lower kind of memory ourselves—that its process is the same in man and dog and squirrel. I am, for instance, riding or walking in a part of the country which all seems unfamiliar, and I have. no recollection of ever having passed that way before; but by-and-by I come to some spot where I have had some little adventure, some mishap, tearing my coat or wounding my hand in getting through a barbed wire fence; or where I had discovered that I had lost something, or left something behind at the inn where I last stayed; or where I had a puncture in my tyre; or where I first saw a rare and beautiful butterfly, or bird, or flower, if I am interested in such things; and the whole scene—the fields and trees, and hedges and farm-house, or cottage below—is all as familiar as possible. But it is the scene that brings back the event. The scene was impressed on the mind at the emotional moment, and is instantly recognised, and at the moment of recognition the associated event is remembered.

CHAPTER VI

THE successive Junes, Julys, and Augusts spent in
this low-lying, warm forest country have served to
restore in my mind the insect world to its proper
place in the scheme of things. In recent years, in
this northern land, it had not seemed so important
a place as at an earlier period of my life in a
country nearer to the sun. Our insects, less numerous, smaller in size, more modest in colouring, and
but rarely seen in swarms and clouds and devastating
multitudes, do not force themselves on our attention,
as is the case in many other regions of the earth.
Here, for instance, where I am writing this chapter,
there is a stretch of flat, green, common land by
the Test, and on this clouded afternoon, at the end

of summer, while sitting on one of the innumerable
little green hillocks covering the common, it seemed
to me that I was in a vacant place where animal
life had ceased to be. Not an insect hummed in
that quiet, still atmosphere, nor could I see one tiny
form on the close-cropped turf at my feet. Yet I
was sitting on one of their populous habitations.
Cutting out a section of the cushion-like turf of
grass and creeping thyme that covered the hill and
made it fragrant, I found the loose, dry earth within
teeming with minute yellow ants, and many of the
hillocks around were occupied by thousands upon
thousands of the same species. Indeed, I calculated
that in a hundred square yards at that spot the
ant inhabitants alone numbered not less than about
two hundred thousand.

It is partly on account of this smallness and secretive-
ness of most of our insects—of our seeing so little of
insect life generally except during the summer heats in
a few favourable localities, and partly an effect of our
indoor life, that we think and care so little about them.
The important part they play, if it is taught us, fades
out of knowledge : we grow in time to regard them as
one of the superfluities in which nature abounds despite
the ancient saying to the contrary. Or worse, as
nothing but pests. What good are they to us indeed!
Very little. The silkworm and the honey-bee have
been in a measure domesticated, and rank with, though
a long way after, our cattle, our animal pets and poultry.

But wild insects! There is the turnip-fly, and the Hessian-fly, and botfly, and all sorts of worrying, and blood-sucking, and disease-carrying flies, in and out of houses; and gnats and midges, and fleas in seaside lodgings, and wasps, and beetles, such as the cockchafer and blackbeetle—are not these all pests? This is the indoor mind—its view of external nature—which makes the society of indoor people unutterably irksome to me, unless (it will be understood) when I meet them in a house, in a town, where they exist in some sort of harmony, however imperfect, with their artificial environment.

I am not concerned now with the question of the place which insects occupy in the scale of being and their part in the natural economy, but solely with their effect on the nature-lover with or without the "curious mind"—in fact, with insects as part of this visible and audible world. Without them, this innumerable company that each "deep in his day's employ" are ever moving swiftly or slowly about me, their multitudinous small voices united into one deep continuous Æolian sound, it would indeed seem as if some mysterious malady or sadness had come upon Nature. Rather would I feel them alive, teasing, stinging, and biting me; rather would I walk in all green and flowery places with a cloud of gnats and midges ever about me. Nor do I wish to write now about insect life generally: my sole aim in this chapter is to bring before the reader some of the most notable species seen in this place—

those which excel in size or beauty, or which for some
other reason are specially attractive. For not only is
this corner of Hampshire most abounding in insect life,
but here, with a few exceptions, the kings and nobles of
the tribe may be met with.

Merely to see these nobler insects as one may see
them here, as objects in the scene, and shining gems in
nature's embroidery, is a delight. And here it may be
remarked that the company of the entomologist is
often quite as distasteful to me out of doors as that of
the indoor-minded person who knows nothing about
insects except that they are a "nuisance." Entomologist generally means collector, and his—the entomologist's—admiration has suffered inevitable decay,
or rather has been starved by the growth of a more
vigorous plant—the desire to possess, and pleasure in
the possession of, dead insect cases.

One summer afternoon I was visiting at a parsonage
in a small New Forest village in this low district, when
my host introduced me to a friend of his, the vicar of a
neighbouring parish, remarking when he did so that I
would be delighted to know him as he was a great
naturalist. The gentleman smiled, and said he was not
a "great naturalist," but only a "Lepidopterist." Now
it happened that just then I had a lovely picture in my
mind, the vivid image of a humming-bird hawk-moth
seen suspended on his misty wings among the tall
flowers in the brilliant August sunshine. I had looked
on it but a little while ago, and thought it one of the

most beautiful things in nature; naturally on meeting a Lepidopterist I told him what I had seen, and something of the feeling the sight had inspired in me. He smiled again, and remarked that the season had not proved a very good one for the Macroglossa stellatarum. He had, so far, seen only three specimens; the first two he had easily secured, as he fortunately had his butterfly net when he saw them. But the third!—he hadn't his net then; he was visiting one of his old women, and was sitting in her garden behind the cottage talking to her when the moth suddenly made its appearance, and began sucking at the flowers within a yard of his chair. He knew that in a few moments it would be gone for ever, but fortunately from long practice, and a natural quickness and dexterity, he could take any insect that came within reach of his hand, however wild and swift it might be. "So!"—the parson Lepidopterist explained, suddenly dashing out his arm, then slowly opening his closed hand to exhibit the imaginary insect he had captured. Well, he got the moth after all! And thus owing to his quickness and dexterity all three specimens had been secured.

I, being no entomologist but only a simple person whose interest and pleasure in insect life the entomologist would regard as quite purposeless—I felt like a little boy who had been sharply rebuked or boxed on the ear. This same Lepidopterist may be dead now, although a couple of summers ago he looked remarkably well and in the prime of life; but I see that some one else is now parson of his parish. I have not taken the pains to inquire; but, dead or alive, I cannot imagine him in that beautiful country of the Future which he perhaps spoke about to the old cottage woman. I cannot imagine him in white raiment, with a golden harp in his hand; for if here, in *this* country, he could see nothing in a humming-bird hawk-moth among the flowers in the sunshine but an object to be collected, what in the name of wonder will he have to harp about!

The humming-bird hawk, owing to its diurnal habits, may be seen by any one at its best; but as to the other species that equal and surpass it in lustre, their beauty, so far as man is concerned, is all wasted on the evening gloom. They appear suddenly, are vaguely seen for a few moments, then vanish; and instead of the clear-cut, beautiful form, the rich and delicate colouring and airy, graceful motions, there is only a dim image of a moving grey or brown something which has passed before us. And some of the very best are not to be seen even as vague shapes and as shadows. What an experience it would be to look on the death's-head moth in a state of nature, feeding among the flowers in the early evening, with some sunlight to show the delicate grey-blue markings and mottlings of the upper- and the inde- scribable yellow of the under-wings — is there in all nature so soft and lovely a hue? Even to see it alive in the only way we are able to do, confined in a box in which we have hatched it from a chrysalis dug up in the potato patch and bought for sixpence from a work-

man, to look on it so and then at its portrait—for artists
and illustrators have been trying to do it these hundred
years—is almost enough to make one hate their art.

My ambition has been to find this moth free, in order
to discover, if possible, whether or no it ever makes its
mysterious squeaking sound when at liberty. But I
have not yet found it, and Lepidopterists I have talked
to on this subject, some who have spent their lives in
districts where the insect is not uncommon, have
assured me that they have never seen, and never expect
to see, a death's-head which has not been artificially
reared. Yet moths there must be, else there would be
no caterpillars and no chrysalis.

One evening, in a potato patch, I witnessed a large
hawk-moth meet his end in a way that greatly surprised
me. I was watching and listening to the shrilling of a
great green grasshopper, or leaf cricket, that delightful
insect about which I shall have to write at some length
in another chapter, when the big moth suddenly
appeared at a distance of a dozen yards from where I
stood. It was about the size of a privet-moth, and had
not been many moments suspended before a spray of
flowers, when a meadow-pipit, which had come there
probably to roost, dashed at and struck it down, and
then on the ground began a curious struggle. The
great moth, looking more than half as big as the
aggressor, beat the pipit with his strong wings in his
efforts to free himself; but the other had clutched the
soft, stout body in its claws, and standing over it with

wings half open and head feathers raised, struck repeatedly at it with the greatest fury until it was killed. Then, in the same savage hawk-like manner, the dead thing was torn up, the pipit swallowing pieces so much too large for it that it had the greatest trouble to get them down. The gentle, timid, little bird had for the moment put on the "rage of the vulture."

In the southern half of the New Forest, that part of the country where insects of all kinds most abound, the moths and butterflies are relatively less important as a feature of the place, and as things of beauty, than some other kinds. The purple emperor is very rarely seen, but the silver-washed fritillary, a handsome, conspicuous insect, is quite common, and when several of these butterflies are seen at one spot playing about the bracken in some open sunlit space in the oak woods, opening their orange-red spotty wings on the broad vivid green fronds they produce a strikingly beautiful effect. It is like a mosaic of minute green tesseræ adorned with red and black butterfly shapes, irregularly placed.

But here the most charming butterfly to my mind is the white admiral, when they are seen in numbers, as in the abundant season of 1901, when the oak woods were full of them. Here is a species which, seen in a collection, is of no more value æsthetically than a dead leaf or a frayed feather dropped in the poultry-yard, or an old postage stamp in an album, without a touch of brilliance on its dull blackish-brown and white wings;

yet which alive pleases the eye more than the splendid and larger kinds solely because of its peculiarly graceful flight. It never flutters, and as it sweeps airily hither and thither, now high as the tree-tops, now close to the earth in the sunny glades and open brambly places in the oak woods, with an occasional stroke of the swift-gliding wings, it gives you the idea of a smaller, swifter more graceful swallow, and sometimes of a curiously-marked, pretty dragon-fly.

When we think of the bright colours of insects, the dragon-flies usually come next to butterflies in the mind, and here in the warmer, well-watered parts of the Forest they are in great force. The noble Anax imperator is not uncommon; but though so great, exceeding all other species in size, and so splendid in his "clear plates of sapphire mail," with great blue eyes, he is surpassed in beauty by a much smaller kind, the *Libellula virgo alis erectus coloratis* of Linnæus, now called Calopteryx virgo. And just as the great imperator is exceeded in beauty by the small virgo, so is he surpassed in that other chief characteristic of all dragon-flies to the unscientific or natural mind, their uncanniness, by another quite common species, a very little less than the imperator in size—the Cordulegaster annulatus.

These names are a burden, and a few words must be said on this point lest the reader should imagine that he has cause to be offended with me personally.

Is it not amazing that these familar, large, showy,

I

and striking-looking insects have no common specific
names with us? The one exception known to me is
the small beautiful virgo just spoken of, and this is
called in books "Demoiselle" and "King George," but
whether these names are used by the people anywhere
or not, I am unable to say. On this point I consulted
an old water-keeper of my acquaintance on the Test.
He has been keeper for a period of forty-six years,
and is supposed to be very intelligent, and to know
everything about the creatures that exist in those waters
and water-meadows. He assured me that he never
heard the names of Demoiselle and King George. "We
calls them dragons and horse-stingers," he said. "And
they do sting, and no mistake, both horse and man."
He then explained that the dragon-fly dashes at its
victim, inflicts its sting, and is gone so swiftly that it
is never detected in the act; but the pain is there, and
sometimes blood is drawn.

Nor had the ancient water-keeper ever heard another
vernacular name given by Moses Harris for this same
species—kingfisher, to wit. Moses Harris, one of our
earliest entomologists, wrote during the last half of
the eighteenth century, but the date of his birth
and the facts of his life are not known. He began to
publish in 1766, his first work being on butterflies and
moths. One wonders if the unforgotten and at-no-time-
neglected Gilbert White never heard of his contem-
porary Moses, and never saw his beautiful illustrations
of British insects, many of which still keep their bright

colours and delicate shadings undimmed by time in his old folios. In one of his later works, *An Exposition of English Insects*, dated 1782, he describes and figures some of our dragon-flies. It was the custom of this author to give the vernacular as well as the scientific names to his species, and in describing the virgo, he says:—" These . . . on account of the brilliancy and richness of the colouring are called kingfishers." But he had no common name for the others, which seemed to trouble him, and at last in desperation after describing a certain species, he says that it is "vulgarly called the dragon-fly!"

I pity old Moses and I pity myself. Why should we have so many suitable and often pretty names for moths and butterflies, mostly small obscure creatures, and none for the well-marked, singular-looking, splendid dragon-flies? The reason is not far to seek. When men in search of a hobby to occupy their leisure time look to find it in some natural history subject, as others find it in postage stamps and a thousand other things, they are, like children, first attracted by those brilliant hues which they see in butterflies. Moreover, these insects when preserved keep their colours, unlike dragon-flies and some others, and look prettiest when arranged with wings spread out in glass cases. Moths being of the same order are included, and so we get the collector of moths and butterflies and the Lepidopterist. So exceedingly popular is this pursuit, and the little creatures collected so much talked and written about, that it has been found convenient to invent English names for them, and thus we have, in moths, wood-tiger, leopard, goat, gipsy, ermine, wood-swift, vapourer, drinker, tippet, lappet, puss, Kentish glory, emperor, frosted green, satin carpet, coronet, marbled beauty, rustic wing and rustic shoulder-knot, golden ear, purple cloud, and numberless others. In fact, one could not capture the obscurest little miller that flutters round a reading-lamp which the Lepidopterist would not be able to find a pretty name for.

The dragon-flies, being no man's hobby, are known only by the old generic English names of dragons, horse-stingers, adder-stingers, and devil's darning-needles. Adder-stinger is one of the commonest names in the New Forest, but it is often simply "adder." One day while walking with a friend on a common near Headley, we asked some boys if there were any adders there. "Oh yes," answered a little fellow, "you will see them by the stream flying up and down over the water." The name does not mean that dragon-flies sting adders, but that, like adders, they are venomous creatures. This very common and wide-spread notion of the insect's evil disposition and injuriousness is due to its shape and appearance—the great fixed eyes, bright and sinister, and the long snake-like plated or scaly body which, when the insect is seized, curls round in such a threatening manner. The colouring, too, may have contributed towards the evil reputation; at all events, one of our largest species has a remarkably serpent-like aspect due to its colour scheme—shining jet black, banded and slashed with wasp-yellow. This is the magnificent Cordulegaster annulatus, little inferior to the Anax imperator in size, and a very common species in the southern part of the New Forest in July. But how astonishing and almost incredible that this singular-looking, splendid, most dragon-like of the dragon-flies should have no English name!

Something remains to be said of the one dragon-fly which has got a name, or names, although these do not appear to be known to the country people. Mr. W. T. Lucas in his useful monograph on the British dragon-flies, writes enthusiastically of this species, Calopteryx virgo, that it is "the most resplendent of our dragon-flies, if not of all British insects." It is too great praise; nevertheless the virgo is very beautiful and curious, the entire insect, wings included, being of an intense deep metallic blue, which glistens as if the insect had been newly dipped in its colour-bath. Unlike other dragon-flies, it flutters on the wing like a butterfly with a weak, uncertain flight, and, again like a butterfly, holds its blue wings erect when at rest. It is one of the commonest as well as the most conspicuous dragon-flies on the Boldre, the Dark Water, and other slow and marshy streams in the southern part of the Forest.

In South America I was accustomed to see

dragon-flies in rushing hordes and clouds, and in masses clinging like swarming bees to the trees; here we see them as single insects, but I once witnessed a beautiful effect produced by a large number of the common turquoise blue dragon-fly gathered at one spot, and this was in Hampshire. I was walking, and after passing a night at a hamlet called Buckhorn Oak, in Alice Holt Forest, I went next morning, on a Sunday, to the nearest church at the small village of Rutledge. It was a very bright windy morning in June, and the oak woods had been stripped of their young foliage by myriads of caterpillars, so that the sunlight fell untempered through the seemingly dead trees on the bracken that covered the ground below. Now, at one spot over an area of about half an acre, the bracken was covered with the common turquoise blue dragon-fly, clinging to the fronds, their heads to the wind, their long bodies all pointing the same way. They were nowhere close together, but very evenly distributed, about three to six inches apart, and the sight of the numberless slips of gem-like blue sprinkled over the billowy vivid green fern was a rare and exceedingly lovely one.

After writing of the lovely haunters of the twilight, and that noblest one of all—

> The great goblin moth who bears
> Between his wings the ruined eyes of death,

and the angel butterfly, and the uncanny dragon-flies —the flying serpents in their splendour—it may seem a great descent to speak of such a thing as a glow-worm, that poor grub-like, wingless, dull-coloured crawler on the ground, as little attractive to the eye as the centipede, or earwig, or the wood-louse which it resembles. Nor is the glow-worm a southern species, since it is no more abundant in the warmest district of Hampshire than in many other parts of the country. Nevertheless, when treating of the Insect Notables of these parts, this species which we call a "worm" cannot be omitted, since it produces a loveliness surpassing that of all other kinds.

Here it may be remarked that all the most beautiful living things, from insect to man, like all the highest productions of human genius, produce in us a sense of the supernatural. If any reader should say in his heart that I am wrong, that it is not so, that he experiences no such feeling, I can but remind him that not all men possess all human senses and faculties. Some of us—many of us—lack this or that sense which others have. I have even met a man who was without the sense of humour. In the case of our "worm," unbeautiful in itself, yet the begetter of so great a beauty, the sense of something outside of nature which shines on us through nature, even as the sun shines in the stained glass of a church window, is more distinctly felt than in the case of any other insect in our country, because of the rarity of such a phenomenon. It is, with us, unique; but many of us know the winged luminous insects of other lands.

Both are beautiful, both mysterious—the winged and the wingless; but one light differs from another in glory even as the stars. The fire-fly is more splendid, more surprising, in its flashes. It flashes and is dark, and we watch, staring at the black darkness, for the succeeding flash. It is like watching for rockets to explode in the dark sky: there is an element of impatience which interferes with the pleasure. To admire and have a perfect satisfaction, the insects must be in numbers, in multitudes, sparkling everywhere in the darkness, so that no regard is paid to any individual light, but they are seen as we see snowflakes.

I fancy that Dante, in describing the appearance of glorified souls in heaven, unless he took it all from Ezekiel, had the fire-fly in his mind—

> From the bosom
> Of that effulgence quivers a sharp flash
> Sudden and frequent in the guise of lightning.

Of all who have attempted to describe and compare the two insects—fire-fly and glow-worm—Thomas Lovell Beddoes is the best. Beddoes himself, in those sudden brilliant letters to his friend Kelsall, of Fareham, in this county, was a sort of human fire-fly. In a letter to Proctor, from Milan, 1824, he wrote: "And what else have I seen? A beautiful and far-famed insect—do not mistake, I mean neither the Emperor, nor the King of Sardinia, but a much finer specimen—the fire-fly. Their bright light is evanescent, and alternates with the darkness, as if the swift whirling of the earth struck fire out of the black atmosphere; as if the winds were being set upon that planetary grindstone, and gave out such momentary sparks from their edges. Their silence is more striking than their flashes, for sudden phenomena are almost invariably attended with some noise, but these little jewels dart along the dark as softly as butterflies. For their light, it is not nearly so beautiful and poetical as our still companion of the dew, the glow-worm, with his drop of moonlight."

I agree with Beddoes, but his pretty description of our insect is not quite accurate, as I saw this evening, when, after copious rain, the sky cleared and a full moon shone on a wet, dusky-green earth. The light of the suspended glow-worm was of an exquisite golden green, and, side by side with it, the moonlight on the wet surface of a polished leaf was shining silver-white.

The light varies greatly in power, according, I suppose, to the degree of excitement of the insect and to the atmospheric conditions. Occasionally you will discover a light at a distance shining with a strange glory, a light which might be mistaken for a will-o'-the-wisp, and on a close view you will probably find that a male is on the scene, and the female, aware of his presence though he may be at some distance from her, invisible in the darkness, has been wrought up to the highest state of excitement. You will find her clinging to a stem or leaf, her luminous part raised, and her whole body swaying in a measured way from side to side. If the insect happens to be a foot or two above the ground, in a tangle of bramble and bracken, with other plants with slender stems and deep-cut leaves, the appearance is singularly beautiful. The light looks as if enclosed within an invisible globe, which may be as much as fifteen inches in diameter, and within its circle the minutest details of the scene are clear to the vision, even to the finest veining of the leaves, the leaves shining a pure translucent green, while outside the mystic globe of light all is in deep shadow and in blackness.

With regard to the attitude of the glow-worm when displaying its light, we see how ignorant of the living creature the illustrators of natural history books have been. In scores of works on our shelves, dating from the eighteenth to the twentieth century, the glow-worm is depicted giving out its light while crawling on the ground, and in many illustrations the male is introduced, and is shown flying down to its mate. They drew their figures not from life, but from specimens in a cabinet, only leaving out the pins. But the glow-worm is not perhaps a very well-known creature. A lady in Hampshire recently asked me if it was a species of mole that came out of its run to exhibit its light in the darkness. The insect invariably climbs up and suspends itself by clinging to a stem or blade or leaf, and the hinder part of the body curls up until its under surface, the luminous

GLOW-WORMS

part, is uppermost, thus making the light visible from the air above. In thick hedges I often find the light four or even five feet above the ground. Occasionally a glow-worm will shine from a flat surface, usually a big leaf on to which it has crawled when climbing. Resting horizontally on the leaf, it curls its abdomen up and over its body after the manner of the earwig, until the light is in the right position.

When we consider these facts—the way in which the body is curved and twisted about in order (as it seems) to exhibit the light to an insect flying through the air above, and the increase in the light when the sexual excitement is at its greatest—the conclusion seems unavoidable that the light has an important use, namely, to attract the male. Unavoidable, I say, and yet I am not wholly convinced. The fire-flies of diurnal habits may be seen flying about, feeding and pairing, by day; yet when evening comes they fly abroad again, exhibiting their light. What the function of the light is, or of what advantage it is to the insect, we do not know. Again, it has seemed to me that the male of the glow-worm, even when attracted to the female, fears the light. Thus, when the excitement of the shining glow-worm has caused me to look for the male, I have found him, not indeed in but outside of the circle of light, keeping close to its borders, moving about on feet and wings in the dark herbage and on the ground. I know very well that not a few observations made by one person, but many — hundreds if possible — by different observers, are needed before we can say positively that the male glow-worm fears or is repelled by the light. But some of my observations make me think that the male of the glow-worm, like the males of many other species in different orders that fly by night, is drawn to the female by the scent, and that the light is a hindrance instead of a help, although in the end he is drawn into it. We always find it exceedingly hard to believe that anything in nature is without a use; but we need not go very far—not farther than our own bodies, to say nothing of our minds—before we are compelled to believe that it is so. We may yet find that the beautiful light of our still companion of the dew is of no more use to it than the precious jewel in the toad's head is to the toad.

The hornet, one of my first favourites, has, to our minds, nothing mysterious like our glow-worm, and nothing serpentine or supernatural in him, but he is a nobler, more powerful, and splendid creature than any dragon-fly. I care not to look at a vulgar wasp nor at any diurnal insect, however fine, when he is by, or his loud, formidable buzzing hum is heard. As he comes out of the oak tree shade and goes swinging by in his shining golden red armature, he is like a being from some other hotter, richer land, thousands of miles away from our cold, white cliffs and grey seas. Speaking of that, our hornet, which is at the head of the family and genus of true wasps in

Britain and Europe, is not only large and splendid for a northern insect, since he is not surpassed in lustre by any of his representatives in other parts of the globe.

I admire and greatly respect him, this last feeling dating back to my experience of wasps during my early life in South America. When a boy I was one summer day in the dining-room at home by myself, when in at the open door flew a grand wasp of a kind I had never seen before, in size and form like the hornet, but its colour was a uniform cornelian red without any yellow. Round the room it flew with a great noise, then dashed against a window-pane, and I, greatly excited and fearing it would be quickly gone if not quickly caught, flew to the window, and dashing out my hand, like the wonderfully clever parson-collector, I grasped it firmly by the back with finger and thumb. Now, I had been accustomed to seize wasps and bees of many kinds in this way without getting stung, but this stranger was not like other wasps, and quickly succeeded in curling his abdomen round, and planting his long sting in the sensitive tip of my forefinger. Never in all my experience of stings had I suffered such pain! I dropped my wasp like the hottest of coals, and saw him fling himself triumphantly out of the room, and never again beheld one of his kind. Even now when I stand and watch English hornets at work on their nests, coming and going, paying no attention to me, a memory of that hornet of a distant land returns to my mind; and it is like a twinge, and I venture on no liberties with Vespa crabro.

The hornet is certainly not an abundant insect, nor very generally distributed. One may spend years in some parts of the country and never see it. I was lately asked by friends in Kent, who have their lonely house in a wooded and perhaps the wildest spot in the county, if the hornet still existed in England, or really was an English insect, as they had not seen one in several years. Now in the woods I frequent in the Forest I see them every day, and the abundance of the hornet is indeed for me one of the attractions of the place. His nests are rarely found in old trees, but are common about habitations, in wood-piles, and old, little-used outhouses. I have heard farmers say in this place that they would not hurt a hornet, but regard it as a blessing. So it is, and so is every insect that helps to keep down the everlasting plague of cattle-worrying and crop-destroying flies and grubs and caterpillars.

But I am speaking of the hornet merely as an Insect Notable, a spot of brilliant colour in the scene, one of the shining beings that inhabit these green mansions. He is magnificent, and it is perhaps partly due to his vivid and lustrous red and gold colour, his noisy flight, and fierce hostile attitudes, and partly to the knowledge of his angry spirit and venomous sting, which makes him look twice as big as he really is.

One of the most impressive sights in insect life is, strange to say, in the autumn, when cold rains and winds and early frosts have already brought to an end all that seemed best and brightest in that fairy world.

This is where an ancient or large ivy grows in some well-sheltered spot on a wall or church, or on large old trees in a wood, and flowers profusely, and when on a warm bright day in late September or in October all the insects which were not wholly dead revive for a season, and are drawn by the ivy's sweetness from all around to that one spot. There are the late butterflies, and wasps and bees of all kinds, and flies of all sizes and colours—green and steel-blue, and grey and black and mottled, in thousands and tens of thousands. They are massed on the clustered blossoms, struggling for a place; the air all about the ivy is swarming with them, flying hither and thither, and the humming sound they produce may be heard fifty yards away like a high wind. One cannot help a feeling of melancholy at this animated scene; but they are anything but melancholy. Their life has been a short and a merry one, and now that it is about to end for ever they will end it merrily, in feasting and revelry.

And never does the hornet look greater, the king and tyrant of its kind, than on these occasions. It swings down among them with a sound that may be heard loud and distinct above the universal hum,

and settles on the flowers, but capriciously, staying
but a moment or two in one place, then moving to
another, the meaner insects all expeditiously making
room for it. And after tasting a few flowers here and
there it takes its departure. These large-sized October
hornets are all females, wanderers from ruined homes,
in search of sheltered places where, foodless and com-
panionless, and in a semi-torpid condition, each may
live through the four dreary months to come. In
March the winter of their discontent will be over, and
they will come forth with the primrose and sweet violet
to be founders and mothers of new colonies—the brave
and splendid hornets of another year; builders, fighters,
and foragers in the green oak-woods; a strenuous hungry
and thirsty people, honey-drinkers, and devourers of
the flesh of naked white grubs and caterpillars, black
and brown and green and gold, and barred and
quaintly-coloured swift aerial flies.

CHAPTER VII

Great and greatest among insects—Our feeling for insect music—
Crickets and grasshoppers—Cicada anglica—Locusta viridissima
—Character of its music—Colony of green grasshoppers—
Harewood Forest—Purple emperor—Grasshoppers' musical
contests—The naturalist mocked—Female viridissima—Over-
elaboration in the male—Habits of female—Wooing of the male
by the female.

I HAD thought to include all or most of the greatest of
the insects known in these parts in the last chapter,
but the hornet, and the vision it called up of that
last revel in the late blossoming ivy on the eve of
winter and cold death seemed to bring that part of
the book to an end. The hornet was the greatest in
the sense that a strong man and conqueror is the
greatest among ourselves, as the lion or wolf among
mammals, and that feathered thunderbolt and scourge,
the peregrine falcon among birds. But there are great
and greatest in other senses; and just as there are
singers, big and little, as well as warriors among the
"insect tribes of human kind," so there are among
these smaller men of the mandibulate division of the
class Insecta. And their singers, when not too loud
and persistent, as they are apt to be in warmer lands
than ours, are among the most agreeable of the inhabitants of the earth. They are less to us than to the

people of the southern countries of Europe—infinitely less than they were to some of the civilised nations of antiquity, and than they are to the Japanese of to-day. This is, I suppose, on account of their rarity with us, for our best singers are certainly somewhat rare or else exceedingly local. The field-cricket, which must be passed over in this chapter to be described later on, is an instance in point. The universal house-cricket is known to and in some degree loved by all or most persons; it is the cricket on the hearth, that warm, bright, social spot when the world outside is dark and cheerless; the lively, companionable sound endears itself to the child, and later in life is dear because of its associations. The field-grasshopper, too, is familiar to every one in the summer pastures; but the best of our insect musicians, the great green grasshopper, appears to be almost unknown to the people. Here, for instance, where I am writing, there is one on the table which stridulates each afternoon, and in the evening when the lamp is lighted. The sustained bright shrilling penetrates to all parts of the house, and in the tap-room of the inn, two rooms away, the villagers, coming in for their evening beer and conversation, are startled at the unfamiliar, sharp, silvery sound, and ask if it is a bird.

Probably it is owing to this rarity of our best insect singers, and partly, too, perhaps to the disagreeable effect on our ears of the loud cicadas heard during our southern travels, that an idea is produced in us of something exotic, or even fantastic, in a taste for insect music. We wonder at the ancient Greeks and the modern Japanese. But it should be borne in mind that the sounds had and have for them an expression they cannot have for us—the expression which comes of association.

If the insects named as our best are rare and local, or at all events not common, what shall we say of our cicada? Can we call him a singer at all? or if he be not silent, as some think, will he ever be more to us than a figure and descriptive passage in a book—a mere cicada of the mind? He is the most local, or has the most limited range, of all, being seldom found out of the New Forest district. He was discovered there about seventy years ago, and Curtis, who gave him the proud name of Cicada anglica, expressed the opinion that he had no song. And many others have thought so too, because, they have been unable to hear him. Others, from Kirby and Spence to our time, have been of a contrary opinion. So the matter stands. A. H. Swinton, in his work on *Insect Variety and Propagation*, 1885, relates that he tried in vain to hear Cicada anglica before going to France and Italy, to make a study of the cicada music; and he writes:—"In northern England their woodland melody has not yet fallen on the ear of the entomologist, but it must not therefore be inferred that these musicians are wholly absent, for among the rich and bounteous southern fauna of Hampshire and Surrey we still retain one outlying waif of the cigales.

. . . *Cicada anglica*, seemingly the montana of Scopoli, if not Hæmatodes in *propriâ personâ*. The male, usually beaten in June from blossoming hawthorn in the New Forest, is provided with instruments of music, and the female, more terrestrial, is often observed wandering with a whirring sound among bracken wastes, where she is thought to deposit her ova."

It struck me some time ago that some of the disappointed entomologists may have heard the sound they were listening for without knowing it. In seeking for an object—some rare little flower, let us say, or a chipped flint, or a mushroom—we set out with an image of it in the mind, and unless the object sought for corresponds to its mental prototype, we in many cases fail to recognise it and pass on. And it is the same with sounds. The listeners perhaps heard a sound so unlike their idea, or image, of a cicada's song, or so like the sound of some other quite different insect, that they paid no attention to it, and so missed what they sought for. At all events, I can say that unless we have some Orthopterous insect of a species unknown to me, which sings in trees, then our cicada does sing, and I have heard it. The sound which I heard, and which was new to me, came from the upper foliage of a large thorn-tree in the New Forest, but unfortunately ·it ceased on my approach, and I failed to find the singer. The entomologist may say that the question remains as it was, but my experience may encourage him to try again. Had I not been expecting to hear an insect singing high up in the trees, I should have said at once that this was a grasshopper's music, though unlike that of any of the species I am accustomed to hear. It was a sustained sound, like that of the great green grasshopper, but not of that excessively bright, subtle, penetrative quality: it was a lower sound, not shrill, and distinctly slower—in other words, the beats or drops of sound which compose the grasshopper's song, and run in a stream, were more distinct and separate, giving it a trilling rather than a reeling character. Had we, in England, possessed a stridulating mantis, which is capable of a slower, softer sound than any grasshopper, I should have concluded that I was listening to one; but there was not, in this New Forest music, the slightest resemblance to the cicada sounds I had heard in former years. The cicadas may be a "merry people," and they certainly had the prettiest things said of them by the poets of Greece, but I do not like their brain-piercing, everlasting whirr; this sound of the English cicada, assuming that I heard that insect, was distinctly pleasing.

But more than cicada, or field-cricket, or any other insect musician in the land, is our great green grass-hopper, or leaf-cricket, Locusta viridissima. I have been accustomed to hear him in July and August, in hedges, gardens, and potato patches at different points along the south coast and at some inland

spots, always in the evening. It is easy, even
after dark, to find him by following up the sound,
when he may be seen moving excitedly about on
the topmost sprays or leaves, pausing at intervals
to stridulate, and occasionally taking short leaps from
spray to spray. He belongs to a family widely distributed on the earth, and in La Plata I was familiar with two species which in form and colour—
a uniform vivid green—were just like our viridissima, but differed in size, one being smaller and the
other twice as large. The smaller species sang by
day, all day long, among water-plants growing in the
water ; the large species stridulated only by night,
chiefly in the maize fields, and was almost as loud
and harsh as the cicadas of the same region. I distinctly remember the sounds emitted by these two
species, and by several other grasshoppers and leaf-
crickets, but none of their sounds came very near
in character to that of viridissima. This is a curious,
and to my sense a very beautiful sound ; and when
a writer describes it as " harsh," which we not un-
frequently find, I must conclude either that one of
us hears wrongly, or not as the world hears, or that,
owing to poverty, he is unable to give a fit expression. It is a sustained sound, a current of brightest,
finest, bell-like strokes or beats, lasting from three or
four to ten or fifteen seconds, to be renewed again
and again after short intervals ; but when the musician is greatly excited, the pauses last only for a

moment—about half a second, and the strain may go on for ten minutes or longer before a break of any length. But the quality is the chief thing; and here we find individual differences, and that some have a lower, weaker note, in which may be detected a buzz, or sibilation, as in the field-grasshopper; but, as a rule, it is of a shrillness and musicalness which is without parallel. The squealings of bats, shrews, and young mice are excessively sharp, and are aptly described as "needles of sound," but they are not musical. The only bird I know which has a note comparable to the viridissima is the lesser white-throat—the excessively sharp, bright sound emitted both as an anger-note and in that low and better song described in a former chapter. It is this musical sharpness which pleases in the insect, and makes it so unlike all other sounds in a world so full of sound. Its incisiveness produces a curious effect: sitting still and listening for some time at a spot where several insects are stridulating, certain nerves throb with the sound until it seems that it is in the brain, and is like that disagreeable condition called "ringing in the ears" made pleasant. Almost too fine and sharp to be described as metallic, perhaps it comes nearer to the familiar sound described by Henley—

> Of ice and glass the tinkle,
> Pellucid, crystal-shrill.

Crystal beads dropped in a stream down a crystal stair would produce a sound somewhat like the insect's song, but duller. We may, indeed, say that this grasshopper's sounding instrument is glass; it is a shining talc-like disc, which may be seen with the unaided sight by raising the elytra.

Some time ago, in glancing through some copies of Newman's monthly *Entomologist*, 1836, I came upon an account of a numerous colony of the great green grasshopper, which the writer found by chance at a spot on the Cornish coast. The effect produced by the stridulating of a large number of these insects was very curious. I envied the old insect-hunter his experience. A colony of viridissima—what a happiness it would be to discover such a thing! And now, late in the summer of 1902, I have found one, and though a very thinly populated one compared to his, it has given me a long-coveted opportunity of watching and listening to the little green people to my heart's content.

The happy spot was in Harewood Forest, a dense oak-wood covering an area of about two thousand acres, a few miles from Andover. I had haunted it for some days, finding little wild life to interest me except the jays, which seemed to be the principal inhabitants. In the middle of this forest or wood, among the oak trees there stands a tall handsome granite cross about thirty feet high, placed to mark the exact spot, known as "Deadman's Plack," where over nine centuries ago King Edgar, with his own hand, slew his friend and favourite, Earl Athelwold. The account which history

gives of this pious monarch, called the Peaceable,
despite his volcanic disposition where women were
concerned, especially his affair with Elfrida, who was
also pious and volcanic as well as beautiful, reads in
these dull, proper times like a tale from another
hotter, fiercer world.   It is not strange that many
persons find their way through the thick forest by
the narrow track to this place or "Plack"; and there
too I went on several days, and sat by the hour and
meditated.   It had struck me as a suitable spot to
watch for the purple emperor; but I saw him not,
and once only I caught sight of his bride to be—a
big black-looking butterfly which rose from the top
of an oak, took a short flight, and returned to settle
once more on the highest leaves in the same place.
This vain hunt for the purple king of the butterflies
—to see him, not to "take"—led to the discovery of
the green minstrels.   Near the cross, or "monument,"
as it is called, there is an open place occupying a
part of the top and a slope of a down, as pretty a
bit of wild heath as may be found in the county.
Stony and barren in places, it is in other parts clothed
in ling, purple with bloom at this season, with a few
pretty little birches and clumps of tangled thorn and
bramble scattered about.   But the feature which gives
a peculiar charm to the spot is the false brome grass
which flourishes on the slope, growing in large patches,
and on the borders of these mixing its vivid light-green
tussocks with the purple-flowered heath.   It is the

species called (in books) heath false brome grass, but as lips of man refuse to pronounce these four ponderous monosyllables, the invention of some dreary botanist, that follow and jolt against each other, I will venture to rename it good-for-nothing grass. For it is useless

The Grasshoppers' Heath

to the farmer, since no domestic herbivore will touch it; its sole justification is its exceeding beauty. It grows as high as a man's knees, or higher, and even in the driest, hottest season keeps its wonderfully vivid fresh green, as near a brilliant colour as any green leaf can be; and the stalks and graceful spikes after the flowering time are pale yellow-brown, and have a golden lustre in the bright August and September sunlight. Could our poetical viridissima have a more suitable home! And here, coming out from the thick oaks and sauntering about the heath I caught the sound of his delicate shrilling, and to my delight found myself in the midst of a colony. They were not abundant, and one could not experience the sensation produced by many stridulating at a time: they were thinly scattered over two or three acres of ground, but at some points I could hear several of them shrilling together at different distances, and it was not difficult to keep two or three in sight at one time.

Hitherto I had known this insect as an evening musician, beginning as a rule after sunset and continuing till about eleven o'clock. Here he made his music only during the daylight hours, from about ten or eleven in the morning until five or six o'clock in the afternoon, becoming silent at noon when it was hot. But it was late in the season when I found him, on August 26, and after much rain the weather had become exceptionally cool for the time of year.

When stridulating it appeared to be the ambition of every male grasshopper to get up as high as he could climb on the stiff blades and thin stalks of the grass; and there, very conspicuous in his uniform green colour which in a strong sunlight looked like the green of verdigris, his translucent overwings glistening like a dragon-fly's wings, he would shrill and make the grass to which he was clinging tremble to his rapidly vibrating body. Then he would listen to the shrill response of some other singer not far off, and then sing and listen again, and yet again; then all at once in a determined manner he would set out to find his rival, travelling high up through the grass, climbing stems and blades until they bent enough for him to grasp others and push on, reminding one of a squirrel progressing through the thin highest branches of a hazel copse. After covering the distance in this manner, with a few short pauses by the way to shrill back an answering challenge, he would find a suitable place near to the other, still in his place high up in the grass; and then the two, a foot or so, sometimes three or four inches, apart, would begin a regular duel in sound at short range Each takes his turn, and when one sings the other raises one of his forelegs to listen; one may say that in lifting a leg he "cocks an ear." The attitude of the insects is admirably given in the accompanying drawing from life. This contest usually ends in a real fight: one advances, and when at a distance of five or six inches makes a leap at his adversary, and the other, prepared for what is coming and in position, leaps too at the same moment, so that they meet midway, and strike each other with their long spiny hind legs. It is done so quickly that the movements cannot be followed by the eye, but that they do hit hard is plain, as in many cases one is knocked down or flung to some distance

RIVAL GREAT GREEN GRASSHOPPERS

away. Thus ends the round; the beaten one rushes off as quickly as he can, as if hurt, but soon pulls up, and lowering his head, begins defiantly stridulating as before. The other follows him up, shrills at and attacks him again; and you may see a dozen or twenty such encounters between the same two in the course of half-an-hour. Occasionally when the blow is struck they grasp each other and fall together; and it is hardly to be doubted that they not only kick, like French wrestlers and bald-headed coots, but also make wicked use of their powerful black teeth. Some of the fighters I examined had lost a portion of one of the forelegs —one had lost portions of two—and these had evidently been bitten off. Perhaps they inflict even worse injuries. Hearing two shrilling against each other at a spot where there was a large clump of heath between them, I dropped down close by to listen and watch, when I discovered a third grasshopper sitting midway between the others in the centre of the heath bush. This one appeared more excited than the others, keeping his wings violently agitated almost without a pause, and yet not the faintest sound proceeded from him. It proved on examination that one of his stiff overwings had been bitten or torn off at the base, so that he had but half of his sounding apparatus left, and no music could his most passionate efforts ever draw from it, and, silent, he was no more in the world of green grasshoppers than a bird with a broken wing in the world of birds.

For it cannot be doubted that his own music is the greatest, the one all-absorbing motive and passion of his little soul. This may seem to be saying too much—to attribute something of human feelings to a creature so immeasurably far removed from us. Fantastic in shape, even among beings invertebrate and unhuman, one that indeed sees with opal eyes set in his green goat-like mask, but who hears with his forelegs, breathes through spiracles set in his sides, whipping the air for other sense impressions and unimaginable sorts of knowledge with his excessively long limber horns, or antennæ, just as a dry fly fisher whips the crystal stream for speckled trout; and, finally, who wears his musical apparatus (his vocal organs) like an electric shield or plaster on the small of his back. Nevertheless it is impossible to watch their actions without regarding them as creatures of like passions with ourselves. The resemblance is most striking when we think not of what we, hard Saxons, are in this cold north, but of the more fiery, music-loving races in warmer countries. I remember in my early years, before the advent of "Progress" in those outlying realms, that the ancient singing contests still flourished among the gauchos of La Plata. They were all lovers of their own peculiar kind of music, singing endless *decimas* and *coplas* in high-pitched nasal tones to the strum-strumming of a guitar; and when any singer of a livelier mind than his fellows had the faculty of improvising, his fame went forth, and the others of his quality were filled with emu-

lation, and journeyed long distances over the lonely
plains to meet and sing against him. How curiously
is this like our island grasshoppers, who have come to
us unchanged from the past, and are neither Saxons
nor Celts, but true, original, ancient Britons—the little
grass-green people with passionate souls! You can
almost hear him say—this little green minstrel you
have been watching when his shrill note has brought
back as shrill an answer—as he resolutely sets out
over the tall, bending grasses in the direction of the
sound, "I'll teach him to sing!"

So interested was I in watching them, so delighted
to be in this society, whose members, for all their
shape, no longer moved about in, to me, unimaginable worlds, that I went day after day and spent
long hours with them. I could best watch their
battles by getting down on my knees in the good-
for-nothing ("heath false brome") grass, so as to
bring my eyes within two or three feet of them.
My attitude, kneeling with bowed head by the half-
hour at a stretch, one day attracted the attention
of some persons who had come in a carriage to pic-
nic under the trees at the foot of the slope, four
or five hundred yards away. There were from time
to time little explosions of laughter, and at last a
young lady of twelve or fourteen cried, or piped out,
in a clear, far-reaching voice, "Holy man!" She
was an impudent monkey.

So far not a word has been said of the female,

simply because, as it seemed to me, there was, so
far, nothing to say. In most insects the odour excites
and draws the males, often from long distances, as
we see in the moths ; they fly to, and find, and see
her, and woo, and chase, and fight with each other
for possession of her ; and when there are beautiful
or fantastic movements, sometimes accompanied with
sounds, corresponding to the antics of birds—I have
observed them in species of Asilidæ and¹ other insects
—they are directly caused by the presence of the
female. But with viridissima it appears not to be
so, since they do not seek the female, nor will
they notice her when she comes in their way, but
they are wholly absorbed in their own music, and in
trying to out-sing the others, or, failing in this, to
kick and bite them into silence.

Now, seeing this strange condition of things among
these insects—seeing it day after day for weeks—the
conclusion forced itself upon my mind that we have
here one of those strange cases among the lower creatures which are not uncommon in human life—the
case of a faculty, a means to an end, being developed
and refined to an excessive degree, and the reflex
effect of this too great refinement on the species, or
race. Comparing it then to certain human matters—
to Art, let us say—we see that that which was but
a means has become an end, and is pursued for its
own sake.

Such a conclusion may seem absurd, and perhaps it

L

is, since we cannot know what "nimble emanations" and vibrations, which touch not our coarser natures, there may be to link these diverse and seemingly ill-fitting actions into one perfect chain. It may be said, for instance, that in this species the incessant stridulating of the male has an action similar to that of the sun's light and heat on plant life, causing the flower to blow and its sexual organs to ripen. But we see, too, that Nature does often overshoot her mark. We have seen it, I think, in the over-refinement of the passion and faculty of fear in certain species, in reference to cases of fascination, and we see it in the over-protected and the over-specialised; but we are so imbued with the idea that the right mean has always been hit upon and adhered to, that it is only in view of the most flagrant cases to the contrary that we are ever startled out of that delusion. The miserable case, for example, of the Polyergus rufescens, the slave-making ant, who, from being too much waited upon, has so entirely lost the power of waiting upon himself that he will perish of hunger amidst plenty if his slaves be not there to pick up and put the food into his mouth. These extreme cases are not the only ones, for every one of such a character there are hundreds of cases. "Degeneration," as Ray Lankester has aptly said, "goes hand in hand with elaboration;" and I would add that in numberless cases over-elaboration is the cause of degeneration.

The female is the grander insect, being nearly a

third larger than the male, of a fuller figure, and adorned with a long, broadsword - shaped ovipositor, which projects beyond her wings like a tail. She has rather a grand air too, and is both silent and inactive. Hers is a life of listening and waiting; and the waiting is long—days and weeks go by, and the males stridulate, and fight, and pay no attention to her. But how patient she can be may be seen in the case of one which I took from her heath and placed on a well-berried branch of wild guelder on my table. There she was contented to rest, usually on one of the topmost clusters, for many days, almost always with the window open at the side of her branch, so that she could easily have made her escape. The wind blew in upon her, and outside the world was green and lit with sunshine. One could almost fancy that she was conscious of her fine appearance in her pale vivid green colour, touched in certain lights with glaucous blue, on her throne of clustered carbuncles. At intervals of an hour or two she would move about a little, and find some other perch; only the waving of her long, fine antennæ appeared to show that she was alive to much that was going on about her—in her world. The one thing that excited her was the stridulating of one of the males confined in a glass vessel on the same table. She would then travel over her branch to get as near as possible to the musician, and would remain motionless, even to the nervous antennæ, and apparently absorbed in the sound for as long as it lasted. At first she ate a few of the crimson berries on her branch, and also took a little parsley and shepherd's purse, but later on she declined all green stuff, and fed on jam, honey, cooked sultanas, and bread-and-butter pudding, which she liked best. Water and ginger-beer for drink. This most placid and dignified lady—we had got into calling her "Lady Greensleeves," and "Queen," and sometimes "The Cow"— was restored, on September 12, in good health, after sixteen days, to her native heath, and disappeared from sight in the long grass, quietly making her way to some spot where she could settle down comfortably to listen to the music.

All the females I found and watched behaved as my captive had done. They were no more active, and preferred to be at a good height above the ground— eighteen inches or two feet—when quietly listening. One day I watched one perched on the topmost spray of a heath bush in her listening attitude: clouds came over the sun, and the wind grew colder and stronger, and the singers ceased singing. And at last, finding that the silence continued, and doubtless feeling uncomfortable on that spray where the wind blew on and swayed her about, she slowly climbed down and settled herself in a horizontal position on the sheltered side of the plant; and when the sun broke out and shone on her she tipped over on one side, stretched her hind legs out, and rested motionless in that position, exactly like a fowl lying in her dusting-place luxuriating in the heat.

But at last despite that air of repose which is her chief characteristic, she is so wrought upon by that perpetual, shrill, irresistible music that she can no longer endure to sit still, but is drawn to it.   She goes to her charmers, one may say, to remind them by her presence that the minstrelsy in which they are so absorbed is not itself an end but a means.   Brisk or lively she cannot be, but it is plain that when she follows up or settles herself down near her forgetful knights, she is greatly excited, and waiting to be taken in marriage.   That she distinguishes one singer above others, or exercises " selection " in the Darwinian sense, seems unlikely: it strikes one, on the contrary, that having so long suffered neglect she is only too willing to be claimed by any one of them.   And this is just what they decline to do—for some time, at any rate.   Again and again I have observed when the female had followed and placed herself close to a couple of these rival musicians, that they took not the least notice of her; and that when, in the course of the alarums and excursions, one of them found himself close to her, the sight of her appeared to disconcert him, and he made all haste to get away from her.   It looked to human eyes as if her large portly figure had not corresponded to his ideal, and had even moved him to repugnance.   But the Ann of Cleves in a green gown is an exceedingly patient person, and very persistent, and though often denied, she will not be denied, or take No for an answer. But it is altogether a curious business, for not only is the wooing process reversed, as many think it is in the cuckoo, but it lasts an unconscionable time in a creature whose life, in the perfect stage, is limited to a season. But the female viridissima has not the power and swiftness of that feathered lady who boldly pursues her singer (in love with nothing but his own voice), and compels him to take her.

in Harewood Forest.

CHAPTER VIII

IN the last chapter I got away—succeeded in breaking
away, would perhaps be a better expression—from that
favourite hunting-ground of mine farther south; and
the reader would perhaps care to know why a book
descriptive of days in Hampshire should be so much
taken up with days in one small corner of the county.
Hampshire is not a very large county compared with
some others: I have traversed it in this and in that
direction often enough to be pretty familiar with a
great deal of it, from the walled-round cornfield which
was once Roman Calleva to the Solent, and from the
beautiful wild Rother on the Sussex border to the Avon
in the west. There is much to see and know within these
limits: for all of those whose proper study is man, his
history and his works; and for the archæologist and for

the artist and seekers after the picturesque, there is much—nay, there is more to attract in the northern than in the southern half of the county. I, not of them, go south, and by preference to one spot, because my chief interest and delight is in life—life in all its forms, from man who walks erect and smiling looks on heaven to the minutest organic atoms—the invisible life. It here comes into my mind that the very smell of the earth, in which we all delight, the smell which fills the air after rain in summer, and is strong when we turn up a spadeful of fresh mould, which the rustic calls "good," believing, perhaps rightly, that we must smell it every day to be well and live long, is after all an odour given off by a living thing—Cladothrix odorifera. Too small for human eyes, which see only objects proportioned to their bigness, so minute, indeed, that millions may inhabit a clod no larger than one's watch, yet are they able to find a passage to us through the other subtler sense; and from the beginning of our earthly journey even to its end we walk with this odour in our nostrils, and love it, and will perhaps take with us a sweet memory of it into the after-life.

Life being more than all else to me, I am drawn to the spot where it exists in greatest abundance and variety.

I remember feeling this passion very strongly one day during this summer of 1902 after looking at a spider. It was an interesting spider, and I found it within a couple of miles of Lyndhurst, of all places;

a spot so disagreeable to me that I avoid it, and look
for nothing and wish for nothing to detain me in its
vicinity.

Lyndhurst is objectionable to me not only because
it is a vulgar suburb, a transcript of Chiswick or
Plumstead in the New Forest where it is in a wrong
atmosphere, but also because it is the spot on which
London vomits out its annual crowd of collectors,
who fill its numerous and ever-increasing brand-new
red-brick lodging-houses, and who swarm through all
the adjacent woods and heaths, men, women, and
children (hateful little prigs!) with their vasculums,
beer and treacle pots, green and blue butterfly nets,
killing bottles, and all the detestable paraphernalia of
what they would probably call "Nature Study."

It happened that one day, a mile or two from Lyndhurst, going along the road I caught sight of a pretty
bit of heath through an opening in the wood, and
turning into it I looked out a spot to rest in, and
was just about to cast myself down when I noticed
a small white spider, disturbed by my step, drop
from a cluster of bell-heath flowers to the ground. I
stood still, and presently the spider, recovered from
its alarm, drew itself up again by an invisible thread
and settled down on the bright-coloured blossoms.
Seating myself close by, I began to watch the strangely
shaped and coloured little creature. It was a Thomisus
—a genus of spiders distinguished by the extraordinary
length of the two pairs of forelegs. The one before

me, Thomisus citreus, is also singular on account of
its colour — pale citron or white — and its habit of
sitting on flowers. This habit and the colour, we
may see, are related. The citreus is not a weaver of
snares, but hunts for its prey, or rather lies in wait
to capture any insect that comes to the flower on
which it sits. On white, yellow, and indeed on most
pale-coloured flowers, it almost becomes invisible. On
the brilliant red bell-heath blossom it showed plainly
enough, but even here it did not look nearly so conspicuous as when on a green leaf.

I had observed this white spider before, but had
always seen it sitting motionless in its flower; this
one was curiously restless, and very soon after I had
settled myself down by its side it began to throw
itself into a variety of strange attitudes. The four
long forelegs would go up all at once and stand out
like rays from the round, white body, and by-and-by
they would drop and hang down like two long strings
from the flower. Pretty soon I discovered the cause
of these actions in the presence of a second spider,
less than half the size of the first, moving about close
by. His smallness and hideling habits had prevented
me from seeing him sooner. This small, active, white
creature was the male, and though moving constantly
about in the heath at a distance of half a foot from
her, it was plain that they could see each other and
also understand each other very well. As he moved
round her, passing by means of the threads he kept

Flower Spiders' Antics

throwing out from spray to spray, she moved round
on her flower to keep him in sight; but though fas-
cinated and drawn to her, he still dreaded, and was
pulled by his fear and his desire in opposite ways.
The excitement of both would increase whenever he
came a little nearer, and their attitudes were then
sometimes very curious, the most singular being one
ot the male when he would raise his body vertically
in the air and stand on his two pairs of forelegs.
When very near, they would extend the long forelegs
and touch one another; but always at this point when
they were closest and the excitement greatest a panic
would seize him, and he would make haste to get to
a safer distance. On two such occasions she, as if
afraid to lose him altogether, quitted her beloved
flower and moved after him, and after wandering
about for some time to no purpose, found another
flower-cluster to settle on. And so the queer wooing
went on, and seemed no nearer to a conclusion, when,
to my surprise, I found that I had been sitting and
lying there, with eyes close to the female spider, for
an hour and a half. Once only, feeling a little bored,
I gently stroked her on the back, which appeared to
please her as much as if she had been a pig and I
had scratched her back with my walking-stick. But
no sooner had the soothing effect passed off than she
began again watching the movements of that fantastic little lover of hers, who loved her for her beautiful white body, but feared her on account of those

poison fangs which he could probably see every time she smiled to encourage him. At the end of my long watch the conclusion of the whole complex business seemed farther off than ever: fear had got the mastery, and the male had put so great a distance between them, and moved now so languidly, that it seemed useless to remain any longer.

I had not been watching alone all this time: when I had been about half-an-hour on the spot I had a visitor, a small miserable-looking New Forest boy; he came walking towards me with a little crooked stick in his hand, and asked me in a low, husky voice if I had seen a pony in that part of the Forest. I told him sharply not to come too near as his steps would disturb a spider I was watching. It did not seem to surprise him that I was there by myself watching a spider, but creeping up he subsided gently on the heath by my side and began watching with me. At intervals when there was a lull in the excitement of the spiders I could spare time for a glance at my poor little companion. He was probably eleven or twelve years old, but his stature was that of a boy of eight—a small, stunted creature, meanly dressed, with light-coloured lustreless hair, pale-blue eyes, and a weary sad expression on his pale face. Yet he called himself a gipsy! But the south of England gipsies are a mixed and degenerate lot. They are now so incessantly harried by the authorities that the best of them settle down in the villages, while those who keep to the old

ways and vagrant open-air life are joined by tramps
and wastrels of every shade of colour. This little
fellow had little or no Romany blood in his watery
veins.

He told me that his people were camping not far
off, and that the party consisted of his parents with
six (the half-dozen youngest) of their thirteen children.
They had a pony and trap; but the pony had got away
during the night, and the father and two or three of
the children were out looking for it in different directions. We talked a little at intervals, and I found
him curiously ignorant concerning the wild life of the
Forest. He assured me that he had never seen the
cuckoo, but he had heard of its singular habits, and
was anxious to know how big a bird it was, also its
colour. In some trees near us a wood-wren was uttering its sorrowful little wailing note of anxiety, and
when I asked him what bird it was, he answered " a
sparrer." Nevertheless he seemed to feel a dim sort
of interest in the spiders we were watching, and at
length our intermittent conversation ceased altogether.
When at last, after a long silence, I spoke, he did not
answer, and glancing round I found that he had gone
to sleep. Lying there with eyes closed, his pale face
on the bright green turf, he looked almost corpse-like.
Even his lips were colourless. Getting up, I placed a
penny piece on the turf beside his little crooked stick,
so that on awaking he should have a gleam of happiness in his poor little soul, and went softly away.

But he was sleeping very soundly, for when after going a couple of hundred yards I looked back he was still lying motionless on the same spot.

But when I looked back, and when, regaining the road, I went on my way, and indeed for long hours after, I saw the boy vaguely, almost like a boy of mist, and was hardly able to recall his features, so faintly had he impressed me; while the spider on her flower, and the small male that wooed and won her many times yet never ventured to take her, were stamped so vividly on my brain, that even if I had wished it I could not have got rid of that persistent image. It made me miserable to think that I had left, thousands of miles away, a world of spiders exceeding in size, variety of shape and beauty and richness of colouring those I found here—surpassing them, too, in the marvellousness of their habits and that ferocity of disposition which is without a parallel in nature. I wished I could drop this burden of years so as to go back to them, to spend half a lifetime in finding out some of their fascinating secrets. Finally, I envied those who in future years will grow up in that green continent, with this passion in their hearts, and have the happiness which I had missed.

I, of course, knew that it was but the too vivid and persistent image of that particular creature on which my attention had been fixed which made me regard spiders generally as the most interesting beings in nature—the proper study of mankind, in fact. But it

is always so; any new aspect, form, or manifestation of
the principle of life, at the moment it comes before the
vision and the mind is, to one who is not a specialist,
attractive beyond all others.

But, after all is said and done, I have as a fact
spent many of my Hampshire days at a distance from
the spots I love best, and my subject in this chapter
will be of my sojourn in that eastern corner of the
county, in the village and parish which all naturalists
love, and which many of them know so well.

It is told in the books that some seventy or eighty
years ago an adventurous naturalist journeyed down
from London by rough ways to the remote village of
Selborne, to see it with his own eyes and describe its
condition to the world. The way is not long nor rough
in these times, and on every summer day, almost
at every hour of the day, strangers from all parts
of the country, with not a few from foreign lands,
may be seen in the old village street. Of these
visitors that come like shadows, so depart, nine in
every ten, or possibly nineteen in every twenty, have
no real interest in Gilbert White and his work and
the village he lived in, but are members of that
innumerable tribe of gadders about the land who
religiously visit every spot which they are told should
be seen.

One morning, while staying at the village, in July
1901, I went at six o'clock for a stroll on the common,

and, on going up the Hanger, noticed a couple of
bicycles lying at the foot of the hill; then, half-way up
I found the cyclists—two young ladies—resting on the
turf by the side of the Zigzag. They were conversing
together as I went by, and one having asked some
question which I did not hear, the other replied, "Oh
no! he lived a very long time ago, and wrote a history
of Selborne. About birds and that." To which the
other returned, "Oh!" and then they talked of something else.

These ladies had probably got up at four o'clock that
morning, and ridden several miles to visit the village
and go up the Hanger before breakfast. Later in the
day they would be at other places where other Hampshire
celebrities, big and little, had been born, or had lived
or died—Wooton St. Lawrence, Chawton, Steventon,
Alresford, Basing, Otterbourne, Buriton, Boldre, and a
dozen more; and one, the informed, would say to her
uninformed companion, "Oh dear, no; he, or she, lived
a long, long time ago, somewhere about the eighteenth
century—or perhaps it was the sixteenth—and did
something, or wrote fiction, or history, or philosophy,
and that." To which the other would intelligently
answer, "Oh!" and then they would remount their
bicycles, and go on to some other place.

Although a large majority of the visitors are of this
description, there are others of a different kind—the
true pilgrims; and these are mostly naturalists who
have been familiar from boyhood with the famous

Letters, who love the memory of Gilbert White, and regard the spot where he was born, to which he was so deeply attached, where his ashes lie, as almost a sacred place.   It is but natural that some of these, who are the true and only Selbornians, albeit they may not call themselves by a name which has been filched from them, should have given an account of a first visit, their impression of a spot familiar in description but never

realised until seen, and of its effect on the mind.   But no one, so far as I know, has told of a second or of any subsequent visit.   There is a good reason for this, for though the place is in itself beautiful and never loses its charm, it is impossible for any one to recover the feeling experienced on a first sight.   If I, unlike others, write of Selborne revisited, it is not because there is anything fresh to say of an old, vanished emotion, a feeling which

M

forms a singular and delightful experience in the life of many a naturalist, and is thereafter a pleasing memory but nothing more.

Selborne is now to me like any other pleasant rural place: in the village street, in the churchyard, by the Lyth and the Bourne, on the Hanger and the Common, I feel that I am—

> In a green and undiscovered ground;

the feeling that the naturalist must or should always experience in all places where nature is, even as Coventry Patmore experienced it in the presence of women. He had paid more than ordinary attention to their ways, and knew that there was yet much to learn.

How irrecoverable the first feeling is—a feeling which may be almost like the sense of an unseen presence, as I have described it in an account of my first visit to Selborne in the concluding chapter in a book on *Birds and Man*—was impressed upon me on the occasion of a second visit two or three years later. There was then no return of the feeling—no faintest trace of it. The village was like any other, only more interesting because of several amusing incidents in bird-life which I by chance witnessed when there. Animals in a state of nature do not often move us to mirth, but on this occasion I was made to laugh several times. At first it was at an owl at Alton. I arrived there in the evening of a wet, rough day in May 1898, too late to walk the five miles that remained to my destination. After securing

a room at the hotel, I hurried out to look at the fine old
church, which Gilbert White admired in his day; but
it was growing dark, so that there was nothing for me
but to stand in the wind and rain in the wet churchyard, and get a general idea of the outline of the
building, with its handsome, shingled spire standing
tall against the wild, gloomy sky. By-and-by a vague
figure appeared out of the clouds, travelling against the
wind towards the spire, and looking more like a ragged
piece of newspaper whirled about the heavens than any
living thing. It was a white owl, and after watching
him for some time I came to the conclusion that he was
trying to get to the vane on the spire. A very idle
ambition it seemed, for although he succeeded again
and again in getting to within a few yards of the point
aimed at, he was on each occasion struck by a fresh
violent gust and driven back to a great distance, often
quite out of sight in the gloom. But presently he
would reappear, still striving to reach the vane. A
crazy bird! but I could not help admiring his pluck,
and greatly wondered what his secret motive in aiming
at that windy perch could be. And at last, after so
many defeats, he succeeded in grasping the metal
cross-bar with his crooked talons. The wind, with all
its fury, could not tear him from it, and after a little
flapping he was able to pull himself up; then, bending down, he deliberately wiped his beak on the bar
and flew away! This, then, had been his powerful,
mysterious motive—just to wipe his beak, which he

could very well have wiped on any branch or barn-roof or fence, and saved himself that tremendous labour!

It was an extreme instance of the tyrannous effect of habit on a wild animal. Doubtless this bird had been accustomed, after devouring his first mouse, to fly to the vane, where he could rest for a few minutes, taking a general view of the place, and wipe his beak at the same time; and the habit had become so strong that he could not forego his visit even on so tempestuous an evening. His beak, if he had wiped it anywhere but on that lofty cross-bar, would not have seemed quite clean.

At Selborne, in the garden at the Wakes, I noticed a pair of pied wagtails busy nest-building in the ivy on the wall. One of the birds flew up to the roof of the house, where, I suppose, he caught sight of a fly in an upper window which looked on to the roof, for all at once he rose up and dashed against the pane with great force; and as the glass pane hit back with equal force, he was thrown on to the tiles under the window. Nothing daunted, he got up and dashed against the glass a second time, with the same result. The action was repeated five times, then the poor baffled bird withdrew from the contest, and, drawing in his head, sat hunched up for two or three minutes perfectly motionless. The volatile creature would not have sat there so quietly if he had not hurt himself rather badly.

One more of the amusing incidents witnessed during

my visit must be told. Several pairs of martins were
making their nests under the eaves of a cottage
opposite to the Queen's Arms, where I stayed; and
on going out about seven o'clock in the morning, I
stood to watch some of the birds getting mud at a
pool which had been made by the night's rain in the
middle of the street. It happened that some fowls
had come out of the inn yard, and were walking or
standing near the puddle picking up gravel or any
small morsel they could find. Among them was a
cockerel, a big, ungainly, yellowish Cochin, in the
hobbledehoy stage of that ugliest and most ungraceful variety. For some time this bird stood idly by
the pool, but by-and-by the movements of the martins
coming and going between the cottage and the puddle
attracted his attention, and he began to watch them
with a strange interest; and then all at once he made
a vicious peck at one occupied in deftly gathering a
pellet of clay close to his great, feathered feet. The
martin flitted lightly away, and after a turn or two,
dropped down again at almost the same spot. The
fowl had watched it, and as soon as it came down
moved a step or two nearer to it with deliberation,
then made a violent dash and peck at it, and was no
nearer to hitting it than before. The same thing
occurred again and again, the martin growing shyer
after each attack; then other martins came, and he,
finding them less cautious than the first, stalked them
in turn and made futile attacks on them. Convinced

at last that it was not possible for him to injure or touch these elusive little creatures, he determined that they should gather no mud at that place, and with head up he watched them circling like great flies around him, dashing savagely at them whenever they came lower, or paused in their flight, or dropped lightly down on the margin. It was a curious and amusing spectacle—the big, shapeless, lumbering bird chasing them round and round the pool in his stupid spite; they by contrast so beautiful in their shining purple mantle, snow-white breast, and stockinged feet, their fairy-like aerial bodies that responded so quickly to every motion of their bright, lively, little minds. It was like a very heavy policeman "moving on" a flock of fairies.

One remembers Æsop's dog in the manger, and thinks that this and many of the apologues are really nothing but everyday incidents in animal life, told just as they happened, with the addition of speech (in some cases quite unnecessary) put in the mouth of the various actors. Æsop's dog did not want to be disturbed in his bed of hay, and was not such an unredeemed curmudgeon as the Selborne fowl; but this unlovely temper or feeling — spite and petty tyranny and persecution—is exceedingly common in the lower animals, from the higher vertebrates down even to the insects.

My third visit to Selborne was in July 1901. I went there on the 12th and stayed till the 23rd. Now

July, when the business of breeding is over or far advanced and all the best songsters are dropping into silence, and when the foliage is deepening to a uniform monotonous dark green, is, next to August, the least interesting month of the year. But at Selborne I was singularly fortunate, although the season was excessively dry and hot. The heat was indeed great all over the country, but I doubt if there exists a warmer village than Selborne, unless it be one in some, to me unknown, coombe in Cornwall or Devon. Thus on July 19th, when the temperature rose to ninety degrees in the shade in the City of London, we had it as high as ninety-four degrees in Selborne. The village lies in a kind of trough at the foot of a wall-like hill. If it were not for the moisture and the greenery that surrounds and almost covers it, hanging, as it were, like a cloud above it, the heat would doubtless have been even greater.

These conditions, in whatever way they may affect the human inhabitants, appear to be exceedingly favourable to the house crickets. It was impossible for any one to walk in the village of an evening without noticing the noise they made. The cottages on both sides of the street seemed to be alive with them, so that, walking, one was assailed by their shrilling in both ears. Hearing them so much sent me in search of their wild cousin of the fields and of the mole cricket, but no sound of them could I hear. It was too late for them to sing. No doubt—

as White conjectured—the artificial conditions which civilised man has made for the house cricket have considerably altered its habits. Like the canary and other finches that thrive in captivity, a uniform indoor climate, with food easily found, have made it a singer all the year round. I trust we shall never take to the Japanese custom of caging insects for the sake of their music; but it is probable that a result of keeping tamed or domesticated field crickets would be to set them singing at all seasons against the cricket on the hearth. A listener would then be able to judge which of the two "sweet and tiny cousins" is the better performer. The house cricket has to my ears a louder, coarser, a more creaky sound; but we hear him, as a rule, in a room, singing, as it were, confined in a big box; and I remember the case of the skylark, and the disagreeable effect of its shrill and harsh spluttering song when heard from a cage hanging against a wall. The field cricket, like the soaring skylark, has the wide expanse of open air to soften and etherealise the sound.

Gilbert White lived in an age which had its own little, firmly - established, conventional ideas about nature, which he, open-air man though he was, did not escape, or else felt bound to respect. Thus, the prolonged, wild, beautiful call of the peacock, the finest sound made by any domesticated creature, was to the convention of the day "disgustful," and as a disgustful sound he sets it down accordingly; and when he speaks of the keen pleasure it gave him to listen to the field cricket, he writes in a somewhat apologetic strain: "Sounds do not always give us pleasure according to their sweetness and melody, nor do harsh sounds always displease. We are more apt to be captivated or disgusted with the associations which they promote than with the notes themselves. Thus the shrilling of the field cricket, though sharp and stridulous, yet marvellously delights some hearers, filling their minds with a train of summer ideas of everything that is rural, verdurous, and joyous."

The delight I know, but I cannot wholly agree with the explanation. A couple of months before this visit to Selborne, on May 25, on passing some small grass fields, enclosed in high, untrimmed hedges, on the border of a pine wood near Hythe, by Southampton Water, I all at once became conscious of a sound, which indeed had been for some considerable time in my ears, increasing in volume as I went on until it forced my attention to it. When I listened, I found myself in a place where field crickets were in extraordinary abundance; there must have been many hundreds within hearing distance, and their delicate shrilling came from the grass and hedges all round me. It was as if all the field crickets in the county had congregated and were holding a grand musical festival at that spot. A dozen or twenty house crickets in a kitchen would have made more noise; this was not loud, nor could it properly

be described as a noise; it was more like a subtle
music without rise or fall or change; or like a continuous, diffused, silvery-bright, musical hum, which
surrounded one like an atmosphere, and at the same
time pervaded and trembled through one like a vibra-
tion. It was certainly very delightful, and the feeling in this instance was not due to association, but,
I think, to the intrinsic beauty of the sound itself.

The Selborne stream, or Bourne, with its meadows
and tangled copses on either side, was my favourite
noonday haunt. The volume of water does not greatly
diminish during the summer months, but in many
places the bed of the stream was quite grown over
with aquatic plants, topped with figwort, huge water-
agrimony, with its masses of powdery, flesh-coloured
blooms, creamy meadow-sweet, and rose-purple loosestrife, and willow-herb with its appetising odour of
codlins and cream. The wild musk, or monkey-flower,
a Hampshire plant about which there will be much
to say in another chapter, was also common. At one
spot a mass of it grew at the foot of a high bank
on the water's edge; from the top of the bank long
branches of briar - rose trailed down, and the rich,
pure yellow mimulus blossoms and ivory-white roses
of the briar were seen together. An even lovelier
effect was produced at another spot by the mingling
of the yellow flowers with the large turquoise-blue
water forget-me-nots.

The most charming of the Selborne wild plants

that flower in July is the musk mallow. It was quite common round the village, and perhaps the finest plant I saw was in the churchyard, growing luxuriantly by a humble grave near the little gate that opens to the Lyth and Bourne. As it is known

Musk Mallow.

to few persons, there must almost every day have been strangers and pilgrims in the churchyard who looked with admiration on that conspicuous plant, with its deep-cut, scented geranium-like, beautiful leaves, tender grey-green in colour, and its profusion of delicate, silky, rose-coloured flowers. Many would look on it as some rare exotic, and wonder at its being there by that lowly green mound. But to the residents it was a musk mallow and nothing more— a weed in the churchyard.

When one morning I found two men mowing the grass, I called their attention to this plant and asked them to spare it, telling them that it was one which the daily visitors to the village would admire above all the red geraniums and other gardeners' flowers which they would have to leave untouched. This simple request appeared to put them out a good deal; they took their hats off and wiped the sweat from their foreheads, and after gravely pondering the matter for some time, said they would "see about it" or "bear it in mind" when they came round to that side. In the afternoon, when the mowing was done, I returned and found that the musk mallow had not been spared.

During my stay I was specially interested in two of the common Selborne birds—the cirl bunting and the swift. At about four o'clock each morning the lively, vigorous song of the cirl bunting would be heard from the gardens or ground of the Wakes, at the foot of the hill. From four to six, at intervals, was his best singing time; later in the day he sang at much longer intervals. There appeared to be three pairs of breeding birds: one at the Wakes, another on the top of the hill to the left of the Zigzag path, and a third below the churchyard. The cock bird of the last pair sang at intervals every day during my visit from a tree in the churchyard, and from a big sycamore growing at the side of it. On July 14 I had a good opportunity of judging the penetrative power of this bunting's voice, for by chance, just as the bells commenced ringing for the six o'clock Sunday evening service, the bird, perched on a small cypress in the churchyard, began to sing. Though only about forty yards from the tower, he was not in the least discomposed by the clanging of the bells, but sang at proper intervals the usual number of times—six or eight—his high, incisive voice sounding distinct through that tempest of jangled metallic music.

I was often at Farringdon, a village close by, and there, too, the churchyard had its cirl bunting, singing merrily at intervals from a perch not above thirty yards from the building. And as at Selborne and Farringdon, so I have found it in most places in Hampshire, especially in the southern half of the county; the cirl is the village bunting whose favourite singing place is in the quiet churchyard or the shade-trees at the farm: compared with other members of the genus he might almost be called our domestic bunting. The yellow-hammer is never heard in a village: at Selborne to find him one had to climb the hill and go out on the common, and there he could be heard drawling out his lazy song all day long. How curious to think that Gilbert White never distinguished between these two species, although it is

probable that he heard the cirl on every summer day
during the greater part of his life.

The swifts at Selborne interested me even more,
and I spent a good many hours observing them; but
the swifts I watched were not, strange to say, the
native Selborne birds. When I arrived I took particular notice of the swallows and swifts—a natural thing
to do in Gilbert White's village. The swallows, I was
sorry to find, had decreased so greatly in numbers
since my former visits that there were but few left.
The house-martins, though still not scarce, had also
fallen off a good deal. Of swifts there were about
eight or nine pairs, all with young in their nests, in
holes under the eaves of different cottages. The old
birds appeared to be very much taken up with feeding their young: they ranged about almost in solitude,
never more than four or five birds being seen together,
and that only in the evening, and even when in company
they were silent and their flight comparatively languid.
This continued from the 12th to the 16th, but on that
day, at a little past seven o'clock in the evening, I was
astonished to see a party of over fifty swifts rushing
through the air over the village in the usual violent
way, uttering excited screams as they streamed by.
Rising to some height in the air, they would scatter
and float above the church for a few moments, then
close and rush down and stream across the Plestor,
coming as low as the roofs of the cottages, then along
the village street for a distance of forty or fifty yards,

after which they would mount up and return to the church, to repeat the same race over the same course again and again. They continued their pastime for an hour or longer, after which the flock began to diminish, and in a short time had quite melted away.

On the following evening I was absent, but some friends staying at the village watched for me, and they reported that the birds appeared after seven o'clock and played about the place for an hour or two, then vanished as before.

On the afternoon of the 18th I went with my friends to the ground behind the churchyard, from which a view of the sky all round can be obtained. Four or five swifts were visible quietly flying about the sky, all wide apart. At six o'clock a little bunch of half-a-dozen swifts formed, and began to chase each other in the usual way, and more birds, singly, and in twos and threes, began to arrive. Some of these were seen coming to the spot from the direction of Alton. Gradually the bunch grew until it was a big crowd numbering seventy to eighty birds, and as it grew the excitement of the birds increased: until eight o'clock they kept up their aerial mad gambols, and then, as on the previous evenings, the flock gradually dispersed.

On the evening of the 19th the performance was repeated, the birds congregated numbering about sixty. On the 20th the number had diminished to about forty, and an equal number returned on the following evening; and this was the last time. We watched in vain for them on the 22nd: no swifts but the half-a-dozen Selborne birds usually to be seen towards evening were visible; nor did they return on any other day up to the 24th, when my visit came to an end.

It is possible, and even probable, that these swifts which came from a distance to hold their evening games at Selborne were birds that had already finished breeding, and were now free to go from home and spend a good deal of time in purely recreative exercises. The curious point is that they should have made choice of this sultry spot for such a purpose. It was, moreover, new to me to find that swifts do sometimes go a distance from home to indulge in such pastimes. I had always thought that the birds seen pursuing each other with screams through the sky at any place were the dwellers and breeders in the locality; and this is probably the idea that most persons have.

I wish I could have visited Selborne again last July, in order to find out whether or not the evening gatherings and pastimes of the swifts occur annually. But I was engaged elsewhere, and at the village I had failed to discover any person with interest enough in such subjects to watch for me. It would have been very strange if I had found such an one.

It was not until October 1902 that I went back, two months after the swifts had gone; but I was well occupied for two or three weeks during this latest visit in observing the ways of a grasshopper.

There has already been much about insects in this book, and it may seem that I am giving a disproportionate amount of space to these negligible atomies; nevertheless I should not like to conclude this chapter without adding an account of yet another species, one indeed worthy to rank among the Insect Notables of Southern England described in a former chapter. The account comes best in this place, since the species had seemed rare, or nowhere abundant, until, in October, I found it most common in Selborne parish; and here I came to know it well, as I had come to know its great green relation, Locusta viridissima, at Longparish. Both are of one family, and are night singers, but the Selborne insect belongs to a different genus—Thamnotrizon—of which it is the only British representative; and in colour and habits it differs widely from the green grasshoppers. The members of this charming family are found in all warm and temperate countries throughout the world: in this island we may say that they are at the extreme northern limit of their range. Of our nine British species only three are found north of the Thames. Thamnotrizon cinereus is one of these, but is mainly a southern species, and the latest of our grasshoppers to come to maturity. In September it is full grown, and may be heard until November. It is much smaller than viridissima, and is very dark in colour, the female, which has no vestige of wings, being of a uniform deep olive-brown, except the under surface, which is bright buttercup-yellow. The male, though

N

smaller than the female, and like her in colour, has a more distinguished appearance on account of his small aborted wings, which serve as an instrument of music, and form a disc of ashy grey colour on his black and brown body.

Unless looked at closely this insect appears black,

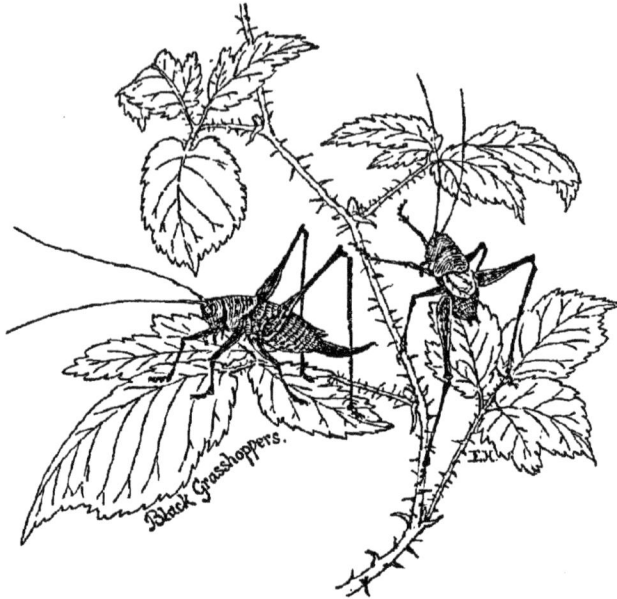

Black Grasshoppers.

and might very well be called the black grasshopper. And here it is necessary once more to protest against what must be regarded as a gross neglect of a plain duty on the part of writers on our native insects who will not give English names even to the most common and interesting species. Unless it has a vernacular name they will go on speaking of it as Thamnotrizon

cinereus, Cordulegaster annulatus, or whatever it may
be, to the end of time.   This grasshopper has no
common name that I can discover: I have caught and
shown it to the country people, asking them to name
it, and they informed me that it was a "grasshopper,"
or else a "cricket."   Black, or black and yellow, or
autumn grasshopper would do very well: but any
English name would be better than the entomologist's
ponderous double name compounded out of two dead
languages.

Our black grasshopper lives in grass and herbage, in
the shade of bushes and trees, and so long as the
weather is hot it is hard to find him, as he keeps in
the shade.   He is furthermore the shyest and wariest
of his family, and ready to vanish on the least alarm.
He does not leap, but slips away into hiding; and
if one goes too near, or attempts to take him, he
suddenly vanishes.   He simply drops down through
the leaves to the earth, and sits close and motionless
at the roots on the dark mould, and unless touched
will not move.   When traced down to his hiding-place
he leaps away, and again sits motionless, where, owing
to his dark colour on the dark soil, he is invisible.
Later, when the weather grows cool, he comes out and
sits on a leaf, basking by the hour in the sun, his
eyes turned from it; and it is then easy to find him,
the dark colour making him appear very conspicuous
on a green leaf.   Occasionally he sings in the afternoon,
but, as a rule, he begins at dusk, and continues for

some hours. To sing, the males often go high up in the bushes, and when emitting their sound are almost constantly on the move.

The sound is a cricket-like chirp; it is never sustained, but in quality it resembles the subtle musical shrilling of the viridissima, although it does not carry half so far.

In disposition the two species, the black and great green grasshoppers, are very unlike. The female viridissima, we have seen, is the most indolent and placid creature imaginable, while the males are perpetually challenging and fighting one another. The males of the black grasshopper I could never detect fighting. It is not easy to observe them, as they sing mostly at night; and as a rule when singing they are well hidden by the leaves. But I have occasionally found two males singing together, apparently against each other, when I would watch them, and although as they moved about they constantly passed and repassed so close that they all but touched, they never struck at each other, nor put themselves into fighting attitudes. One day I found two males sitting on a leaf together, side by side, like the best of friends, basking in the sun.

The female, on the other hand, is a most unpleasant creature, so restless that in confinement she spends the whole time in running about in her cage or box, incessantly trying to get out, examining everything, eating of everything given her, and persecuting any other insect placed with her. When I put males and females together the poor males were kicked and bitten until they died.

Before visiting Selborne in October, it had seemed to me that hunting for this grasshopper was a most fascinating pursuit. It was very hard to find him by day, and when by chance you caught sight of him, sitting on a green leaf in the sun and looking like a small, very dark-coloured frog with abnormally long hind legs, it was generally in a bramble bush, into which he would vanish when approached too near.

When at Selborne, one evening I heard one singing among the herbage at the foot of the Hanger, and next morning I found one at the same spot—a female, sitting on a gold-red' fallen beech leaf, her blackness on the brilliant leaf making her very conspicuous. A little later, when the wet weather improved, I found the grasshopper all about the village, and even in it; but it was most abundant near the Well Head and in the hedges between Selborne and Nore Hill. Here on a sunny morning I could find a score or more of them, and at dark they could be heard in numbers chirping in all the hedges.

CHAPTER IX

It is a pleasure to be at Selborne; nevertheless I
find I always like Selborne best when I am out of
it, especially when I am rambling about that bit of
beautiful country on the border of which it lies.
The memory of Gilbert White; the old church with
its low, square tower and its famous yew tree; above
all, the constant sight of the Hanger clothed in its
beechen woods—green, or bronze and red-gold, or
purple-brown in leafless winter—all these things do
not prevent a sense of lassitude, of ill-being, which
I experience in the village when I am too long in it,
and which vanishes when I quit it, and seem to breathe
a better air. This is no mere fancy, nor something
peculiar to myself; the natives, too, are subject to this
secret trouble, and are, some of them, conscious of it.
Round about Selborne you will find those who were
born and bred in the village, who say they were never

well until they quitted it; and some of these declare
that they would not return even if some generous
person were to offer them a cottage rent free. The
appearance of the people, too, may be considered
in this connection. Mary Russell Mitford exclaims in
one of her village sketches that there was not a pretty
face in the country side. The want of comeliness
which is so noticeable in the southern parts of Berkshire is not confined to that county. The people of
Berkshire and Hampshire, of the blonde type, are very
much alike. But there are degrees; and if you want
to see, I will not say a handsome, nor a pretty, but
a passably fresh and pleasant face among the cottagers,
you must go out of Selborne to some neighbouring
village to look for it.

But this question does not now concern us. The
best of Selborne is the common on the hill—all the
better for the steep hill which must be climbed to
get to it, since that difficult way prevents the people
from making too free use of it, and regarding it as a
sort of back-yard or waste place to throw their rubbish
on. It is a perpetual joy to the children. One morning in October I met there some youngsters gathering
kindling wood, and feasting at the same time on wild
fruits—the sloes were just then at their best. They
told me that they had only recently come to live in
Selborne from Farringdon, their native village. "And
which place do you like best?" I asked. "Selborne!"
they shouted in a breath, and indeed appeared sur-

prised that I had asked such a question. No wonder. This hill-top common is the most forest-like, the wildest in England, and the most beautiful as well, both in its trees and tangles of all kinds of wild plants that flourish in waste places, and in the prospects which one gets of the surrounding country. Here, seeing the happiness of the boys, I have wished to be a boy again. But one does not think so much of this spot when one comes to know the country round, and finds that Selborne hill is but one of many hills of the same singular and beautiful type, sloping away gently on one side, and presenting a bold, almost precipitous front on the other, in most cases clothed on the steep side with dense beech woods. It is now eight years since I began to form an acquaintance with this east corner of Hampshire, but not until last October (1902) did I know how beautiful it was. From Selborne Hill one sees something of it; a better sight is obtained from Nore Hill, where one is able to get some idea of the peculiar character of the scenery. It is all wildly irregular, high and low grounds thrown together in a pretty confusion, and the soil everywhere fertile, so that the general effect is of extreme richness. One sees, too, that the human population is sparse, and that it has always been as it is now, and man's work—his old irregular fields, and the unkept hedges which, like the thickets on the waste places, are self-planted, and have been self-planted for centuries, and the old deep-winding lanes and by-roads—have come at last to seem one with nature's work. Out of this broken, variegated, richly green surface, here and there, in a sort of range, but irregular like all else, the hills, or hangers, lift their steep, bank-like fronts — splendid masses of red and russet gold against the soft grey-blue autumnal sky.

It is delightful to walk through this bit of country from Nore Hill, and from hill to hill, across green fields, for the farms are here like wild lands that all are free to use, to Wheatham Hill, the highest point, which rises 800 feet above the sea-level. From this elevation one looks over a great part of that green variegated country of the Hangers, and sees on one hand where it fades close by into the sand and pine district beginning at Wolmer Forest, and on another side, beyond the little town of Petersfield, the region of great rolling downs stretching far away into Sussex.

In my rambles about this corner of Hampshire, during which I visited all the villages nearest to Selborne—Empshott, Hawkley, Greatham, East and West Tisted, Worldham, Priors Dean, Colemore, Privett, Froxfield, Hartley Maudit, Blackmore, Oakhanger, Kingsley, Farringdon, and Newton Valence—I could not help thinking a good deal about Hampshire village churches generally. It was a subject which had often enough been in my mind before in other parts of the county, but it now came back to me in connection with Gilbert White's strictures on these sacred buildings. Their "meanness" produced a feeling in him which is the nearest approach to indignation discoverable in his pages. He is speaking of jackdaws breeding in rabbit holes, and shrewdly conjectures that this habit has arisen on account of the absence of steeples and towers suitable as nesting-places. "Many Hamp-

Prior's Dean.

shire places of worship," he remarks, "make no better appearance than dovecotes." He envied Northamptonshire, Cambridgeshire, the Fens of Lincolnshire, and other districts, the number of spires which presented themselves in every point of view, and concludes: "As an admirer of prospects I have reason to lament this want in my own county, for such objects are very necessary ingredients in an elegant landscape."

The honoured historian of the parish of Selborne
makes me shudder in this passage. But I am, perhaps,
giving too much importance to his words, since one
may judge from his mention of Norfolk in this connection as being even worse off than his own county,
that he was not well informed on the subject. Norfolk, like Somerset, abounds in grand old churches
of the Perpendicular period. That smallness, or
"meanness" as he expresses it, of the Hampshire
churches is, to my mind, one of their greatest merits.
The Hampshire village would not possess that charm
which we find in it—its sweet rusticity and homeliness,
and its harmonious appearance in the midst of a nature
green and soft and beautiful—but for that essential
feature and part of it, the church which does not tower
vast and conspicuous as a gigantic asylum or manufactory from among lowly cottages dwarfed by its
proximity to the appearance of pigmy-built huts in the
Aruwhimi forest. These immense churches which in
recent years have lifted their tall spires and towers
amidst lowly surroundings in many rural places, are, as
a rule, the work of some zealot who has seared his sense
of beauty with a hot iron, or else of a new over-rich lord
of the manor, who must have all things new, including
à big new church to worship a new God in—his own
peculiar Stock Exchange God, who is a respecter of
wealthy persons. Here in Hampshire we have seen the
old but well preserved village church pulled down—
doubtless with the consent of the ecclesiastical authori-

ties—its ancient monuments broken up and carted away, its brasses made into fire ornaments by cottagers or sold as old metal, and the very gravestones used in paving the scullery and offices of the grand new parsonage built to match the grand new church.

When coming upon one of these "necessary ingredients in an elegant landscape" in some rural spot I have sometimes wondered what the feeling of the people who have spent their lives there can be about it. What effect has the new vast building, with its highly decorated yet cold and vacant interior, on their dim minds—on their religion, let us say? It may be a poor unspiritual sort of religion, based on old traditions and associations, mostly local; but shall we scorn it on that account? If we look a little closely into the matter, we see that all men, even the most intellectual, the most spiritual, are subject to this feeling in some degree, that it is in all religions. That which from use, from association, becomes symbolic of faith is in itself sacred. At the present time the Church is torn with dissensions because of this very question. Certain bodily positions and signs and gestures, and woven fabrics and garments of many patterns and colours, and wood and stone and metal objects, and lighted candles and perfumes—mere hay and stubble to others who have different symbols—are things essential to worship in some. Touch these things and you hurt their souls; you deprive them of their means of communication with another world. So the poor peasant who was born and lives in a thatched cottage, with his limited intelligence, his animism, associates the idea of the unseen world with the sacred objects he has seen and known and handled—the small ancient building, the red-barked, dark-leafed yew, the green mounds and lichened gravestones among which he played as a child, and the dim, low-roofed interior of what was to him God's House. Whatever there is in his mind that is least earthly, whatever thoughts he may have of the unseen world and a life beyond this life, were inseparably bound up with these visible things.

We need not follow this line any farther; those who believe with me that the sense of the beautiful is God's best gift to the human soul will see that I have put the matter on other and higher grounds. The small village church with its low tower or grey-shingled spire among the shade trees, is beautiful chiefly because man and nature with its softening processes have combined to make it a fit part of the scene, a building which looks as natural and harmonious as an old hedge which man planted once and nature replanted many times, and as many an old thatched timbered cottage, and many an old grey ruin, ivy grown, with red valerian blooming on its walls.

To pull down one of these churches to put in its place a gigantic Gothic structure in brick or stone, better suited in size (and ugliness) for a London or Liverpool church than for a small rustic village in Hampshire, is nothing less than a crime.

When calling to mind the churches known to me in this part of Hampshire, I always think with peculiar pleasure of the smaller ones, and perhaps with the most pleasure of the smallest of all—Priors Dean.

It happened that the maps which I use in my Hampshire rambles and which I always considered the best —Bartholomew's two miles to the inch—did not mark Priors Dean, so that I had to go and find it for myself. I went with a friend one excessively hot day in July, by Empshott and Hawkley through deep by-roads so deep and narrow and roofed over with branches as to seem in places like tunnels. On that hot day in the silent time of the year it was strangely still, and gave one the feeling of being in a country long deserted by man. Its only inhabitants now appeared to be the bullfinches. In these deep shaded lanes one constantly hears the faint plaintive little piping sound, the almost inaudible alarm note of the concealed bird; and at intervals, following the sound, he suddenly dashes out, showing his sharp-winged shape and clear grey and black upper plumage marked with white for a moment or two before vanishing once more in the overhanging foliage.

We went a long way round, but at last coming to an open spot we saw two cottages and two women and a boy standing talking by a gate, and of these people we asked the way to Priors Dean. They could not tell us. They knew it was not far away—a mile perhaps; but they had never been to it, nor seen it, and didn't well know the direction. The boy when asked shook his head. . A middle-aged man was digging about thirty yards away, and to him one of the women now called, "Can you tell them the way to Priors Dean?"

The man left off digging, straightened himself, and gazed steadily at us for some moments. He was one of the usual type—nine in every ten farm labourers in this corner of Hampshire are of it—thinnish, of medium height, a pale, parchment-face, rather large straightish nose, pale eyes with little speculation in them, shaved mouth and chin, and small side whiskers as our fathers wore them. The moustache has not yet been adopted by these conservatives. The one change they have made is, alas! in their dress — the rusty black coat for the smock frock.

When he had had his long gaze, he said, "Priors Dean?"

"Yes, Priors Dean," repeated the woman, raising her voice.

He turned up two spadefuls of earth, then asked again, "Priors Dean?"

"Priors Dean!" shouted the woman. "Can't you tell 'em how to get to it?" Then she laughed. She had perhaps come from some other part of the country where minds are not quite so slow, and where the slow-minded person is treated as being deaf and shouted at.

Then, at last, he stuck his spade into the soil, and leaving it, slowly advanced to the gate and told us to follow a path which he pointed out, and when we got on the hill we would see Priors Dean before us.

And that was how we found it. There is a satirical saying in the other villages that if you want to find the church at Priors Dean you must first cut down the

Priors Dean Farm house.

nettles. There were no nettles nor weeds of any kind, only the small ancient church with its little shingled spire standing in the middle of a large green graveyard with about a dozen or fifteen gravestones scattered about, three old tombs, and, close to the building, an ancient yew tree. This is a big, and has been a bigger, tree, as a large part of the trunk has perished on one side, but as it stands it measures nearly twenty-four feet round a yard from the earth. This, with a small farmhouse, in old times a manor house, and its out-buildings and a cottage or two, make the village. So quiet a spot is it that to see a human form or hear a human voice comes almost as a surprise. The little antique church, the few stones, the dark ancient tree—these are everything, and the effect on the mind is strangely grateful—a sense of enduring peace, with something of that solitariness and desolation which we find in unspoilt wildernesses.

From these smallest churches, which appear like a natural growth where they are seen, I turn to the large and new, and the largest of all at this place—that of Privett. From its gorgeous yet vacant and cold interior, and from the whole vast structure, including that necessary ingredient in an elegant landscape, the soaring spire visible for many miles around, I turn away as from a jarring and discordant thing — the feeling one experiences at the sight of those brand-new big houses built by over-rich stock jobbers on many hills and open heaths in Surrey and, alas! in Hampshire.

I do not, however, say that all new and large churches raised in small rustic centres appear as discordant things. Even in the group of villages which I have named there is a new and comparatively large one which moves one to admiration—the church of Black-moor. Here the vegetation and surroundings are unlike those which accord best with the small typical structures, the low tower and shingled spire. The tall, square tower of Blackmoor, of white stone roofed with red tiles, rises amid the pines of Wolmer Forest, simple and beautiful in shape, and gives a touch of grace and grateful colour to that darker, austere nature. From every point of view it is a pleasure to the eye, and because of its enduring beauty the memory of the man who raised it is like a perfume in the wilderness.

It is, however, time that bestows the best grace, the indescribable charm to the village church—long centuries of time, which gives the feeling, the expression, of immemorial peace to the weathered and ivied building itself and the surrounding space, the churchyard, with its green heaps, and scattered stones, and funeral yew.

The associated feeling, the *expression*, is undoubtedly the chief thing in the general effect, but the constituents or objects which compose the scene are in themselves pleasing; and one scarcely less important than the building itself, the universal grass, the dark, red-barked tree, is the gravestone. I mean the gravestone that is attractive in shape, which may be seen in every old village churchyard in Hampshire; for not all the stones are of this character. The stone that is beautiful dates back half a century at least, but very few are as old as a century and a half. When we get that far and farther back the inscription is obliterated or indecipherable. Only here and there we may by chance find some stone, half buried in the soil, of an exceptional hardness, marking the-spot where lieth one who departed this life in the seventeenth or early in the eighteenth century. There are many old stones, it is true, with nothing legible on them, but one does not know how old they are.   It is not that these gravestones are beautiful only because they are old, and have had their hard surface softened and embroidered with green moss and lichen of many shades, from pale-grey to orange and red and brown.   The form of the stone, the stone-cutter's work, was beautiful before Nature began to work on it with her sunshine, her rain, her invisible seed.   I cannot think why this old fashion, or rather, let us say, this tender, sacred custom, of marking the last resting-place of the dead with a memorial satisfying to the æsthetic sense, should have gone or died out. The gravestones used at the present time are, as a rule, twice as big as the old ones, and are perfectly plain— immense stone slabs, inscribed with big, fat, black letters differing in size, the whole inscription curiously resembling the local auctioneer's bills to be seen pasted up on barn-doors, fences, and other suitable places. So big, and hard, and bold, and ugly—I try not to see them!

Look from these at the old stone which the earth-worms have been busy trying to bury for a century, until the lower half of the inscription is under ground; the stone which the lichen has embossed and richly coloured; round which the grass grows so close and lovingly, and the small creeping ivy tries to cover. This which has been added to it is but a part of its beauty: you see that its lines are graceful, that they were made so; that the inscription—"Here lyeth the body," &c.—is not cut in letters in use in newspapers and advertising placards, and have therefore no common nor degrading associations, but are letters of other forms, graceful, too, in their lines; and that above the inscription there are sculptured and symbolic figures and lines—emblems of mortality, eternal hope, and a future life—heads of cherubs, winged and blowing on horns, and the sun and wings; skulls and crossbones, and hour-glass and scythe; the funeral urn and weeping-willow; the lighted torch; the heart in flames, or bleeding, or transfixed with arrows; the angel's trumpet, the crown of glory, the palm and the lily, the laurel leaf, and many more.

Did we think this art, or this custom, too little a thing to cherish any longer? I cannot find any person with a word to say about it. I have tried and the result was curious. I have invited persons of my acquaintance into an old churchyard and begged them to look on this stone and on this—the hard ugliness of one, an insult to the dead, and the beauty, the pathos, of the other. And they have immediately fallen into a melancholy silence, or else they have suddenly become angry, apparently for no cause. But the reason probably was that they had never given a thought to the subject, that when they had buried

some one dear to them—a mother or wife or daughter
—they simply went to the stonemason and ordered a
gravestone, leaving him to fashion it in his own way.
The reason of the reason—the full explanation of the
singular fact that they, in these house-beautiful and
generally art-worshipping times, had given no thought
to the matter until it was unexpectedly sprung upon
them; and that if they had lived, say, a hundred years
ago, they would have given it some thought—this the
reader will easily find out for himself.

It is comforting to reflect that gravestones do not
last for ever, nor for very long; and in the meantime
Nature is doing what she can with our ugly modern
memorials, touching, softening, and tingeing them with
her mosses, lichens, and with algæ—her beautiful
iolithus. In most churchyards in southern England
we see many stones stained a peculiar colour, a bright
rust red, darkest in dry weather, and brightest in wet
summers, often varying to pink and purple and orange;
but whatever the hue or shade the effect on the grey
stone, lichened or not, is always beautiful. It is not
a lichen; when the staining is looked closely at nothing is seen but a roughness, a powdery appearance,
on the stone's surface. It is an aerial alga of the
genus Croöleptus, confined to the southern half of
England, and most common in Hampshire, where its
beautifying blush may sometimes be seen on old stone
walls of churches, and old houses and ruins; but it

flourishes most on gravestones, especially in moist situations. The stone must not be too hard, and must, moreover, be acted on by the weather for well-nigh half a century before the alga begins to show on it; but you will sometimes see it on an

Here Lyeth ye body
of Ann wife of Joseph

Old Gravestones

exceptionally soft stone dating no more than thirty or forty years back. On old stones it is very common and peculiarly beautiful in wet summers. In June 1902, after many days of rain, I stood one evening at the little gate at Brockenhurst churchyard, and counted between me and the church twenty gravestones stained with the red alga, showing a richness and variety of

colouring never seen before, the result of so much wet
weather. For this alga, which plays so important a part
in nature's softening and beautifying effect on man's
work—which is mentioned in no book unless it be some
purely technical treatise dealing with the lower vegetable forms—this alga, despite its aerial habit, is still in
essence a water-plant: the sun and dry wind burn its
life out and darken it to the colour of ironstone, so that
to any one who may notice the dark stain it seems a
colour of the stone itself; but when rain falls the colour
freshens and brightens as if the old grey stone had
miraculously been made to live.

If never a word has been written about that red
colour with which Nature touches the old stones to
make them beautiful, a thousand or ten thousand
things have been said about the yew, the chief
feature and ornament of the village churchyard, and
many conjectures have we seen as to the reason
of the very ancient custom of planting this tree
where the dead are laid. The tree itself gives a
better reason than any contained in books. It says
something to the soul in man which the talking or
chattering yew omitted to tell the modern poet; but
very long ago some one said, in the *Death of Fergus*,
"Patriarch of long-lasting woods is the yew; sacred
to forests as is well known." That ancient sacred
character, which survived the introduction of Chris-
tianity, lives still in every mind that has kept any

vestige of animism, the root and essence of all that
is wonderful and sacred in nature.　That red and
purple bark is the very colour of life, and this tree's
life, compared with other things, is everlasting.　The
stones we set up as memorials grow worn and seamed
and hoary with age, even like men, and crumble to
dust at last; in time new stones are put in their
place, and these, too, grow old and perish, and are
succeeded by others; and through all changes, through
the ages, the tree lives on unchanged.　With its
huge, tough, red trunk; its vast, knotted arms out-
stretched; its rich, dark mantle of undying foliage,
it stands like a protecting god on the earth, patriarch and monarch of woods; and indeed it seems but
right and natural that not to oak nor holly, nor any
other reverenced tree, but to the yew it was given to
keep guard over the bodies and souls of those who
have been laid in the earth.

The yew is sometimes called the "Hampshire weed,"
on account of its abundance in the county; if it must
have a second name, I suggest that the Hampshire
or British dragon-tree would be a better and more
worthy one.　It would admirably fit some ancient
churchyard yews in the neighbourhood of Selborne,
especially that of Farringdon.

In the great mass of literature concerning Gilbert
White, there is curiously little said about this village;
yet it has one of the most interesting old churches
in the county—the church in which White offici-

ated for over a quarter of a century, during all the
best years of his life, in fact; for when he resigned
the curacy at Farringdon to take that of Selborne,
for which he had waited so long, he was within two
years of bidding a formal farewell to natural history,
and within eight of his death.   The church register
from 1760 to 1785 is written in his clear, beautiful
hand, and in the rectory garden there is a large
Spanish chestnut-tree planted by him.   Although not
so fortunate in its surroundings as Selborne, with
its Lyth and flowery Bourne and wooded Hanger,
Farringdon village with its noble church and fine
old farm - buildings and old cottages, is the better
village of the two.   At the side of the churchyard
there is an old oast-house, now used as a barn, which
for quaintness and beauty has hardly its match in
England.   The churchyard itself is a pretty, peaceful
wilderness, deep in grass, with ivy and bramble hanging to the trees, and spreading over tombs and mounds.
Long may it be kept sacred from the gardener, with
his abhorred pruning-hook, his basket of geranium
cuttings—inharmonious flower!—and his brushwood
broom to make it all tidy.   Finally, there is the
wonderful old yew.·

A great deal has been written first and last about
the Selborne yew, which appears to rank as one of
the half-dozen biggest yew trees in the country.   Its
age is doubtless very great, and may greatly exceed
the "thousand years" usually given to a very large

churchyard yew. The yews planted two hundred years ago by Gilbert White's grandfather in the parsonage garden close by, are but saplings in comparison. A black poplar would grow a bigger trunk in less than ten years. The Selborne yew was indeed one of the antiquities of the village when White described it a century and a quarter ago. It is, moreover, the best-grown, healthiest, and most vigorous-looking yew of its size in Britain. The Farringdon yew, the bigger tree, has a far more aged aspect — the appearance of a tree which has been decaying for an exceedingly long period.

Trees, like men, have their middle period, when their increase slowly lessens until it ceases altogether; their long stationary period, and their long decline· each of these periods may, in the case of the yew, extend to centuries; and we know that behind them all there may have been centuries of slow growth. The Selborne yew has added something to its girth since it was measured by White, and is now twenty-seven feet round in its biggest part, and exceeds by at least three feet the big yew at Priors Dean, and the biggest of the three churchyard yews at Hawkley. The Farringdon yew in its biggest part, about five feet from the ground, measures thirty feet, and to judge by its ruinous condition it must have ceased adding to its bulk more than a century ago. One regrets that White gave no account of its size and appearance in his day. It has, in the usual manner,

decayed above and below, the upper branches dying
down while the trunk rots away beneath, the tree
meanwhile keeping itself alive and renewing its youth,
as it were, by means of that power which the yew
possesses of saving portions of its trunk from complete decay by covering them inside and out with
new bark.

In the churchyard yew at Crowhurst, Surrey, we
see that the upper part of the tree has decayed until
nothing but the low trunk, crowned with a poor fringe
of late branches, has been left; in this case the trunk
remains outwardly almost entire—an empty shell or
cylinder, large enough to accommodate fourteen persons
on the circular bench placed within the cavity. In
other cases we see that the trunk has been eaten
through and through, and split up into strips; that
the strips, covered inside with new bark, have become
separate trunks, in some instances united above, as in
that of the yew in South Hayling churchyard. The
Farringdon tree has decayed below in this way; long
strips from the top to the roots have rotted and turned
to dust; and the sound portions, covered in and out
with bark, form a group of half-a-dozen flattened boles,
placed in a circle, all but one, which springs from the
middle, and forms a fantastically twisted column in the
centre of the edifice. Between this central strangely
shaped bole, now dead, and the surrounding ring there
is space for a man to walk round in.

It is a wonderful tree, which White looked at every

day for five and twenty years, yet never mentioned, and which Loe says nothing about in his *Yew Trees of Great Britain and Ireland.* The title of this work is misleading : *Famous Yew Trees* it should have been, since it is nothing but a collection of facts as to size, supposed age, &c., of trees that have often been measured and described, and are accordingly well known. It is well, to my way of thinking, that he attempted nothing more. It is always a depressing thought, when one has discovered a wonderful or a beautiful thing, that a very full and very exact account of it is and must be contained in some musty monograph by some industrious, dreary person. At all events, I can say that the yew trees which have most attracted me, which come up when I think of the yew as a wonderful and a sacred tree, are not in the book. Of my Hampshire favourites I will, for a special reason, speak of but one more—the yew in the churchyard of Hurstbourne Priors, a small village on the upper Test, near Andover.

This tree, which is doubtless very aged, has not grown an enormous trunk, nor is it high for an old yew, but its appearance is nevertheless strangely impressive, owing to the length of its lower horizontal branches, which extend to a distance of thirty to thirty-five feet from the trunk, and would lie on the ground if not kept up by props. Another thing which makes one wonder is the number of graves that are crowded together beneath these vast sheltering arms. One may count over thirty stones, some very old; many more have probably perished, and there are besides many green mounds. I have watched in a churchyard in the Midlands a grave being dug under a yew, at about three yards' distance from the trunk: a barrowful of roots was taken out during the process. It seemed to me that a very serious injury was being inflicted on the tree, and it is probable that many of our very old

churchyard yews have been dwarfed in their growth by such cutting of the roots. But what shall we say of the Hurstbourne Priors yew, from which not one but thirty or forty barrow-loads of living roots must have been taken at various times to make room for so many coffins! And what is the secret of the custom in this, and probably other villages, of putting the dead so close to or under the shelter of the tree?

Compare this Hurstbourne Priors yew, and many

other ancient churchyard yews in Hampshire, with that of Selborne, which albeit probably no older is double their size: is it not probable that the Selborne tree is the largest, best grown, and most vigorous of the old yews because it has not been mutilated at its roots as the others have been?

There is but one grave beneath or near this tree; not the grave of any important person, but a nameless green mound of some obscure peasant. I had often looked with a feeling almost of astonishment at that solitary conspicuous mound in such a place, midway between the trunk of the tree and the church door, wondering who it was whose poor remains had been so honoured, and why it was. Then by chance I found out the whole story; but it came to me in scraps, at different times and places, and that is how I will give it to the reader, in fragments, in the course of the following chapter.

CHAPTER X

THE first part of the story of that Selborne mound in a strange place was heard at Wolmer Forest, over five years ago, during my first prolonged visit to that spot. I have often been there since, and have stayed many days, but a first impression of a place, as of a face, is always the best, the brightest, the truest, and I wish to describe Wolmer as I saw it then.

It struck me on that visit that the pleasure we have in visible nature depends in a measure on contrast and novelty. Never is moist verdure so refreshing and delightful to the eye as when we come to it from brown heaths and grey barren downs and uplands. So, too, the greenness of the green earth sharpens our pleasure in all stony and waste places; trim flower gardens show us the beauty of thorns and briars, and make us in love with desolation. As in light and shade, wet and dry, tempest and calm, so the peculiar attractions of each scene and aspect of nature are best "illustrated by their contraries."

I had, accordingly, the best preparation for a visit
to Wolmer by a few days' ramble in Alice Holt Forest,
with its endless oaks, and in the luxuriant meadows
and cool shady woods at Waverley Abbey. It was a
great change to Wolmer Forest. Although its soil is
a "hungry, bare sand," it has long been transformed
from the naked heath of Gilbert White's time to a
vast unbroken plantation. Looked upon from some
eminence it has a rough, dark aspect. There are no
smooth summits and open pleasant places; all is covered
by the shaggy mantle of the pines. But it is nowhere
gloomy, as pine woods are apt to be: the trees are
not big enough, on account of that hungry sand in
which they are rooted, or because they are not yet
very old. The pines not being too high and shady to
keep the sun and air out, the old aboriginal vegetation
has not been killed: in most places the ling forms a
thick undergrowth, and looks green, while outside of
the forest, in the full glare of the sun, it has a harsh,
dry, dead appearance.

On account of this abundance of ling a strange and
lovely appearance is produced in some favourable years,
when the flowers are in great profusion and all the
plants blossom at one time. That most beautiful sight
of the early spring, when the bloom of the wild hyacinth
forms a sheet of azure colour under the woodland trees,
is here repeated in July, but with a difference of hue
both in the trees above and in the bloom beneath.

In May, Wolmer is comparatively flowerless, and

there is no bright colour, except that of the earth itself in some naked spot. The water of the sluggish boggy streamlets in the forest, tributaries of the well-named Dead Water, takes a deep red or orange hue from the colour of the soil. The sand abounds with ironstone, which in the mass is deep rust-red and purple coloured. When crushed and pulverised by traffic and weather on the roads, it turns to a vivid chrome yellow. In the hot noonday sun the straight road that runs through the forest appeared like a yellow band or ribbon. That was a curious and novel picture, which I often had before me during the excessively dry and windy weather in May—the vast whity-blue, hot sky, without speck or stain of cloud above, and the dark forest covering the earth, cut through by that yellow zone, extending straight away until it was lost in the hazy distance. Even stranger was the appearance when the wind blew strongest and raised clouds of dust from the road, which flew like fiery yellow vapours athwart the black pines.

In a small house by the roadside in the middle of the forest I found a temporary home. My aged landlady proved a great talker, and treated me to a good deal of Hampshire dialect. Her mind was well stored with ancient memories. At first I let her ramble on without paying too much attention; but at length, while speaking of the many little ups and downs of her not uneventful life, she asked me if I knew Selborne, and then informed me that she was a native of that village, and

P

that her family had lived there for generations. Her
mother had reached the age of eighty-six years; she
had married her third husband when over seventy. By
her first she had had two and by her second thirteen
children, and my informant, who is now aged seventy-
six, was the last born. This wonderful mother of hers,
who had survived three husbands, and whose memory
went back several years into the eighteenth century,
had remembered the Rev. Gilbert White very well:
she was aged about twelve when he died. It was
wonderful, she said, how many interesting things she
used to tell about him; for Gilbert White, whose name
was known to the great world outside of his parish,
was often in her mind when she recalled her early
years. Unfortunately, these interesting things had
now all slipped out of my landlady's memory. Whenever I brought her to the point she would stand with
eyes cast down, the fingers of her right hand on her
forehead, trying—trying to recall something to tell
me: a simple creature, who was without imagination,
and could invent nothing. Then little by little she
would drift off into something else—to recollections
of people and events not so remote in time, scenes
she had witnessed herself, and which had made a
deeper impression on her mind. One was how her
father, her mother's second husband, had acted as
horn-blower to the "Selborne mob," when the poor
villagers were starving; and how, blowing on his
horn, he had assembled his fellow-revolutionists, and

led them to an attack on the poorhouse, where they
broke down the doors and made a bonfire of the
furniture; then on to the neighbouring village of
Headley to get recruits for their little army. Then
the soldiery arrived on the scene, and took them
prisoners and sent them to Winchester, where they
were tried by some little unremembered Judge Jeffreys,
who sentenced many or most of them to transportation;
but not the horn-blower, who had escaped, and was in
hiding among the beeches of the famous Selborne
Hanger. Only at midnight he would steal down
into the village to get a bite of food and hear the
news from his vigorous and vigilant wife. At length
during one of these midnight descents, he was seen,
and captured, and sent to Winchester. But by this
time the authorities had grown sick—possibly ashamed
—of dealing so harshly with a few poor peasants,
whose sufferings had made them mad, and the horn-
blower was pardoned, and died in bed at home when
his time came.

I did not cease questioning the poor woman because
she would not admit that all she had heard about
Gilbert White was gone past recall. Often and often
had she thought of what her mother had told her.
Up to within two or three years ago she remembered
it all so well. What was it now? Once more, standing dejected in the middle of the room, she would
cudgel her old brains. So much had happened since
she was a girl. She had been brought up to farm-

work. Here would follow the names of various farms in the parishes of Selborne, Newton Valence, and Oak-hanger, where she had worked, mostly in the fields; and of the farmers, long dead and gone most of them, who had employed her. All her life she had worked hard, struggling to live. When people complained of hard times now, of the little that was paid them for their work, she and her husband remembered what it was thirty and forty and fifty years ago, and they wondered what people really wanted. Cheap food, cheap clothing, cheap education for the children—everything was cheap now, and the pay more. And she had had so many children to bring up—ten; and seven of them were married, and were now having so many children of their own that she could hardly keep count of them.

It was idle to listen; and at last, in desperation, I would jump up and rush out, for the wind was calling in the pines, and the birds were calling, and what they had to tell was just then of more interest than any human story.

Not far from my cottage there was a hill, from the summit of which the whole area of the forest was visible, and the country all round for many leagues beyond it. I did not like this hill, and refused to pay it a second visit. The extent of country it revealed made the forest appear too small; it spoilt the illusion of a practically endless wilderness, where I could stroll about all day and see no cultivated spot, and no house, and perhaps no human form. It was, moreover, positively disagreeable to be stared at across the ocean of pines by a big, brand-new, red-brick mansion, standing conspicuous, unashamed, affronting nature, on some wide heath or lonely hillside.

A second hill, not far from the first, was preferable when I wished for a wide horizon, or to drink the wind and the music of the wind. Round and dome-like, it stood alone; and although not so high as its neighbour, it was more conspicuous, and seen from a distance appeared to be vastly higher. The reason of this was that it was crowned with a grove of Scotch firs with boles that rose straight and smooth and mast-like to a height of about eighty feet; thus, seen from afar, the hill looked about a hundred feet higher than it actually was, the tree-tops themselves forming a thick, round dome, conspicuous above the surrounding forest, and Wolmer's most prominent feature. I have often said of Hampshire—very many persons have said the same —that it lacks one thing—sublimity, or, let us say, grandeur. I have been over all its high, open down country, and upon all its highest hills, which, although rising to a thousand feet above the sea at one point, yet do not impress one so much as the South Downs; and I have been in all its forest lands, which have wildness and a thousand beauties, and one asks for nothing better. But the Hollywater Clump in Wolmer Forest as soon as I come in sight of it wakes in me another sense and feeling; and I have found in conversation

with others on this subject that they are affected in
the same way. I doubt if any one can fail to experience such a feeling when looking on that great hill-top
grove, a stupendous pillared temple, with its dome-like
black roof against the sky, standing high above and
dominating the sombre pine and heath country for
miles around.

Gilbert White described Wolmer as a naked heath
with very few trees growing on it. The Hollywater
Clump must, one cannot but think, have been planted
before or during his time. One old native of Wolmer
whose memory over five years ago went back about
sixty years, assured me that the trees looked just as big
when he was a little boy as they do now. Undoubtedly
they are very old, and many, we see, are decaying, and
some are dead, and for many years past they have
been dying and falling.

The green woodpecker had discovered the unsound-
ness of many of them, in some of the trunks, in their
higher part, the birds had made several holes. These
were in line, one above the other, like stops in a flute.
Most of these far-up houses or flats were tenanted by
starlings. This was only too apparent, for the starling,
although neat and glossy in his dress, is an untidy
tenant, and smears the trunk beneath the entrance to
his nest with numberless droppings. You might fancy
that he had set himself to whitewash his tenement,
and had carelessly capsized his little bucket of lime
on the threshold.

HOLLYWATER CLUMP: WOLMER FOREST

.

It was pleasant in the late afternoon to sit at the feet
of these stately red columns—this brave company of
trees, that are warred against by all the winds of
heaven, and look upon the black legions of the forest
covering the earth beneath them for miles.   High
up in the swaying, singing tops a kind of musical
talk was audible—the starlings' medley of clinking,
chattering, wood-sawing, knife-grinding, whistling, and
bell-like sounds.   Higher still, above the tree-tops, the
jackdaws were at their aerial gambols, calling to one
another, exulting in the wind.   They were not breeding
there, but were attracted to the spot by the height of
the hill, with its crown of soaring trees.   Some strong-
flying birds—buzzards, kites, vultures, gulls, and many
others—love to take their exercise far from earth,
making a playground of the vast void heaven.   The
wind-loving jackdaw, even in his freest, gladdest
moments, never wholly breaks away from the earth,
and for a playground prefers some high, steep place—a
hill, cliff, spire, or tower—where he can perch at intervals, and from which he can launch himself, as the
impulse takes him, either to soar and float above, or
to cast himself down into the airy gulf below.

Stray herons, too,. come to the trees to roost.   The
great bird could be seen far off, battling with the wind,
rising and falling, blown to this side and that, now displaying his pale under-surface, and now the slaty blue
of his broad, slow-flapping wings.

As the sun sank nearer to the horizon, the tall

trunks would catch the level beams and shine like fiery
pillars, and the roof thus upheld would look darker and
gloomier by contrast. With the passing of that red
light, the lively bird-notes would cease, the trees would
give forth a more solemn, sea-like sound, and the day
would end.

My days, during all the time I spent at Wolmer,
when I had given up asking questions, and my poor
old woman had ceased cudgelling her brains for lost
memories, were spent with the birds. The yaffle, nightjar, and turtle-dove were the most characteristic species.
Wolmer is indeed the metropolis of the turtle-doves,
even as Savernake is (or was) that of the jays and jackdaws. All day long the woods were full of the low,
pleasing sound of their cooing: as one walked among
the pines they constantly rose up in small flocks from
the ground with noisy wings, and as they flew out into
some open space to vanish again in the dark foliage,
their wings in the strong sunlight often looked white as
silver. But the only native species I wish now to speak
of is the teal as I found it a little over five years ago.
In Wolmer these pretty entertaining little ducks have
bred uninterruptedly for centuries, but I greatly fear
that the changes now in progress—the increase of the
population, building, the larger number of troops kept
close by, and perhaps, too, the slow drying up of the
marshy pools—will cause them to forsake their ancient
haunts.

By chance I very soon discovered their choicest

breeding-place, not far from that dome-shaped, fir-crowned hill which was my principal landmark. This was a boggy place, thirty or forty acres in extent, surrounded by trees and overgrown with marsh weeds and grasses, and in places with rushes. Cotton grass grew in the drier parts, and the tufts nodding in the wind looked at a distance like silvery white flowers. At one end of the marsh there were clumps of willow and alder, where the reed-bunting was breeding and the grasshopper warbler uttered his continuous whirring sound, which seemed to accord with the singing of the wind in the pines. At the other end there was open water with patches of rushes growing in it; and here at the water's edge, shaded by a small fir, I composed myself on a bed of heather to watch the birds.

The inquisitive moor-hens were the first to appear, uttering from time to time their sharp, loud protest. Their suspicion lessened by degrees, but was never wholly laid aside; and one bird, slyly leaving the water, made a wide circuit and approached me through the trees in order to get a better view of me. A sudden movement on my part, when he was only three yards from me, gave him a terrible fright. Mallards showed them-selves at intervals, swimming into the open water, or rising a few yards above the rushes, then dropping out of sight again. Where the rushes grew thin and scattered, ducklings appeared, swimming one behind the other, busily engaged in snatching insects from the surface. By-and-by a pair of teal rose up, flew straight towards me, and dropped into the open water within eighteen yards of where I sat. They were greatly excited, and no sooner touched the water than they began calling loudly; then, from various points, others rose and hurried to join them, and in a few moments there were eleven, all disporting themselves on the water at that short distance. Teal are always tamer than ducks of other kinds, but the tameness of these Wolmer birds was astonishing and very delightful. For a few moments I imagined they were excited at my presence, but it very soon appeared that they were entirely absorbed in their own affairs and cared nothing about me. What a wonderfully lively, passionate, variable, and even ridiculous little creature the teal is! Compared with his great relations—swans, geese, and the bigger ducks—he is like a monkey or squirrel among stately bovine animals. Now the teal have a world-wide range, being found in all climates, and are of many species; they are, moreover, variable in plumage, some species having an exceedingly rich and beautiful colouring; but wherever found, and however different in colour, they are much the same in disposition—they are loquacious, excitable, violent in their affections beyond other ducks, and, albeit highly intelligent, more fearless than other birds habitually persecuted by man. A sedate teal is as rare as a sober-coloured humming-bird. The teal is also of so social a temper that even in the height of the breeding season he is accustomed to meet his fellows at little gatherings. A curious thing is that at these meetings they do not, like most social birds, fall into one mind, and comport themselves in an orderly, disciplined manner, all being moved by one contagious impulse. On the contrary, each bird appears to have an impulse of his own, and to follow it without regard to what his fellows may be doing. One must have his bath, another his frolic; one falls to courting, another to quarrelling, or even fighting, and so on, and the result is a lively splashing, confused performance, which is amusing to see. It was an exhibition of this kind which I was so fortunate as to witness at the Wolmer pond. The body-jerking antics and rich, varied plumage of the drakes gave them a singular as well as a beautiful appearance; and as they dashed and splashed about, sometimes not more than fourteen yards from me, their motions were accompanied by all the cries and calls they have—their loud call, which is a bright and lively sound; chatterings and little, sharp, exclamatory notes; a long trill, somewhat metallic or bell-like; and a sharp, nasal cry, rapidly reiterated several times, like a laugh.

After they had worked off their excitement and finished their fun they broke up into pairs and threes, and went off in various directions, and I saw no more of them.

It was not until the sun had set that a snipe appeared. First one rose from the marsh and began to play over it in the usual manner, then another rose to keep him company, and finally a third. Most of the time they hovered with their breasts towards me, and seen through my glass against the pale luminous sky, their round, stout bodies, long bills, and short, rapidly vibrating wings, gave them the appearance of gigantic insects rather than birds.

At length, tired of watching them, I stretched myself out in the ling, but continued listening, and while thus occupied an amusing incident occurred. A flock of eighteen mallards rose up with a startled cry from the marsh at a distance, and after flying once or twice round, dropped down again. Then the sound of crackling branches and of voices talking became audible advancing round the marsh towards me. It was the first human sound I had heard that day at that spot. Then the sounds ceased, and after a couple of minutes of silence I glanced round in the direction they had proceeded from, and beheld a curious sight. Three boys, one about twelve years old, the others smaller, were grouped together on the edge of the pool, gazing fixedly across the water at me. They had taken me for a corpse, or an escaped criminal, or some such dreadful object, lying there in the depth of the forest. The biggest boy had dropped on to one knee among the rough heather, while the others, standing on either side, were resting their hands on his shoulders. Seen thus, in their loose, threadbare, grey clothes and caps, struck motionless, their white, scared faces, parted lips, and wildly staring eyes turned to me, they were like a group cut in stone. I laughed and waved my hand to them, whereupon their faces relaxed, and they immediately dropped into natural attitudes. Very soon they moved away among the trees, but after eight or ten minutes they reappeared near me, and finally, from motives of curiosity, came uninvited to my side. They proved to be very good specimens of the boy naturalist; thorough little out-laws, with keen senses, and the passion for wildness strong in them. They told me that when they went bird-nesting they made a day of it, taking bread and cheese in their pockets, and not returning till the evening. For an hour we talked in the fading light of day on the wild creatures in the forest, until we could no longer endure the cloud of gnats that had gathered round us.

About three years after the visit to Wolmer I made the acquaintance of a native of Selborne, whose father had taken part as a lad in the famous " Selborne mob," and who confirmed the story I had heard about the horn-blower, whose name was Newland. He had been a soldier in his early manhood before he returned to his native village and married the widow who bore him so many children. It was quite true that he had died at home, in bed, and what was more, he added, he was buried just between the church porch and the yew, where he was all by himself. How he came to be buried there he did not know.

Lately, in October 1902, I heard the finish of the

story. I found an old woman, a widow named Garnett, an elder sister of the woman at Wolmer Forest. She is eighty years old, but was not born until a year or two after the "Selborne mob" events, which fixes the date of that outbreak about the year 1820. She has a brother, now in a workhouse, about two years older than herself, who was a babe in arms at that time. When Newland was at last captured and sent to Winchester, his poor wife, with her baby in her arms, set out on foot to visit him in gaol. It was a long tramp for her thus burdened, and it was also in the depth of one of the coldest winters ever known. She started early, but did not get to her destination until the following morning, and not without suffering a fresh misfortune by the way. Before dawn, when the cold was most intense, while walking over Winchester Hill, her baby's nose was frozen; and though everything proper was done when she arrived at the houses, it never got quite right. His injured nose, which turns to a dark-blue colour and causes him great suffering in cold weather, has been a trouble and misery to him all his life long.

Newland, we know, was forgiven and returned to spend the rest of his life in his village, where he died at last of sheer old age, passing very quietly away after receiving the sacrament from the vicar, and in the presence of his faithful old wife and his children and grandchildren.

After he was dead, two of his children—my informant,

and that brother who as a babe had travelled to Winchester in his mother's arms in cold weather—talked together about him and his life, and of all he had suffered and of his goodness, and in both their minds there was one idea, an anxious wish that his descendants should not allow him to go out of memory. And there was no way known to them to keep him in mind except by burying him in some spot by himself, where his mound would be alone and apart. Finally, brother and sister, plucking up courage, went to the vicar, the well-remembered Mr. Parsons, who built the new vicarage and the church school, and begged him to let them bury their father by the yew tree near the porch, and he good-naturedly consented.

That was how Newland came to be buried at that spot; but before many days the vicar went to them in a great state of mind, and said that he had made a terrible mistake, that he had done wrong in consenting to the grave being made there, and that their father must be taken up and placed at some other spot in the churchyard. They were grieved at this, but could say nothing. But for some reason the removal never took place, and in time the son and daughter themselves began to regret that they had buried their father there where they could never keep the mound green and fresh. People going in or coming out of church on dark evenings stumbled or kicked their boots against it, or when they stood there talking to each other they would rest a foot on it, and romping children sat on it, so that it always had a ragged, unkept appearance, do what they would.

It is certainly an unsightly mound. It would be better to do away with it, and to substitute a small memorial stone with a suitable inscription placed level with the turf.

CHAPTER XI

THE history of the horn-blower and his old wife, and their still living aged children, serves to remind me that this book, which contains so much about all sorts of creatures and forms of life, from spiders and flies to birds and beasts, and from red alga on gravestones to oaks and yews, has so far had almost nothing to say about our own species—of that variety which inhabits Hampshire.

If the critical reader asks what is here meant by "variety," what should I answer him? On going directly from any other district in southern England to the central parts of Hampshire one is sensible of a difference in the people. One is still in southern England, and the peasantry, like the atmosphere, climate, soil, the quiet but verdurous and varied scenery, are more or less like those in other neighbouring counties

—Surrey, Sussex, Kent, Berkshire, Wilts, and Dorset. In general appearance, at all events, the people are much the same; and the dialect, where any survives, and even the quality of the voices, closely resemble those in adjoining counties. Nevertheless there is a difference; even the hasty seers who are almost without the faculty of observation are vaguely cognisant of it, though they would not be able to say what it consisted in. Probably it would puzzle any one to say wherein Hampshire differed from all the counties named, since each has something individual; therefore it would be better to compare Hampshire with some one county near it, or with a group of neighbouring counties in which some family resemblance is traceable. Somerset, Devon, Wilts, and Dorset—these answer the description, and I leave out Cornwall only because its people are unknown to me. The four named have seemed to me the most interesting counties in southern England, but if I were to make them five by adding Hampshire, the verdict of nine persons out of ten, all equally well acquainted with the five, would probably be that it was the least interesting. They would probably say that the people of Hampshire were less good-looking, that they had less red colour in their skins, less pure colour in their eyes; that they had less energy, if not less intelligence, or at all events were less lively, and had less humour.

These differences between the inhabitants of neighbouring and of adjoining counties are doubtless in some measure due to local conditions, of soil, climate, food, customs, and so on, acting for long generations on a stay-at-home people: but the main differences are undoubtedly racial; and here we are on a subject in which we poor ordinary folk who want to know are like sheep wandering shepherdless in some wilderness,

bleating in vain for guidance in a maze of fleece-tearing brambles. It is true that the ethnologists and anthropologists triumphantly point out that the Jute type of man may be recognised in the Isle of Wight, and in a less degree even in the Meon district; for the rest, with a wave of the hand to indicate the northern half of the county, they say that all that is or ought to be more or less Anglo-Saxon. That's all;

since, as they tell us, the affinities of the South Hampshire people, of the New Forest district especially, have not yet been worked out. Not being an anthropologist I can't help them; and am even inclined to think that they have left undone some of the things which they ought to have done. The complaint was made in a former chapter that we had no monograph on fleas to help us; it may be made, too, with regard to the human race in Hampshire. The most that one can do in such a case, since man cannot be excluded from the subjects which concern the naturalist, is to record one's own poor little unscientific observations, and let them go for what they are worth.

There is little profit in looking at the towns'-people. The big coast towns have a population quite as heterogeneous as that of the metropolis; even in a comparatively small rural inland town, like Winchester, one would be puzzled to say what the chief characteristics of the people were. You may feel in a vague way that they are unlike the people of, say, Guildford, or Canterbury, or Reading, or Dorchester, but the variety in forms and faces is too great to allow of any definite idea. The only time when the people even in a town can be studied to advantage in places like Winchester, Andover, &c., is on a market day, or on a Saturday afternoon, when the villagers come in to do their marketing. I have said, in writing of Somerset and its people, that the gentry, the landowners, and the wealthy residents generally, are always in a sense foreigners. The man may bear a name which has been for many generations in a county, but he is never racially one with the peasant; and, as John Bright once said, it is the people who live in cottages that make the nation. His parents and his grand-parents and his ancestors for centuries have been mixing their blood with the blood of outsiders. It is well always to bear this in mind, and in the market-place or the High Street of the country town to see the carriage people, the gentry, and the important ones generally as though one saw them not, or saw them as shadows, and to fix the attention on those who in face and carriage and dress proclaim themselves true natives and children of the soil.

Even so there will be variety enough—a little more perhaps than is wanted by the methodic mind anxious to classify these "insect tribes." But after a time— a few months or a few years, let us say—the observer will perceive that the majority of the people are divisible into four fairly distinct types, the minority being composed of intermediate forms and of nonde-scripts. There is an enormous disproportion in the actual numbers of the people of these distinct types, and it varies greatly in different parts of the county. Of the Hampshire people it may be said generally, as we say of the whole nation, that there are two types— the blonde and the dark; but in this part of England there are districts where a larger proportion of dark blood than is common in England generally has produced a well-marked intermediate type; and this is one of my four distinct Hampshire types. I should place it second in importance, although it comes a very long way after the first type, which is distinctly blonde.

This first most prevalent type, which greatly out-numbers all the others put together, and probably includes more than half of the entire population, is strongest in the north, and extends across the county from Sussex to Wiltshire. The Hampshire people in that district are hardly to be distinguished from those of Berkshire. One can see this best by looking at the school-children in a number of North Hampshire and Berkshire villages. In sixty or seventy to a hundred and fifty children in a village school you will seldom find as many as a dozen with dark eyes.

As was said in a former chapter, there is very little beauty or good looks in this people; on the other hand, there is just as little downright ugliness; they are mostly on a rather monotonous level, just passable in form and features, but with an almost entire absence of any brightness, physical or mental. Take the best-looking woman of this most common type—the de-scription will fit a dozen in any village. She is of medium height, and has a slightly oval face (which, being Anglo-Saxon, she ought not to have), with fairly good features; a nose fairly straight, or slightly aquiline, and not small; mouth well moulded, but the lips too thin; chin frequently pointed. Her hair is invariably

brown, without any red or chestnut colour in it, generally of a dull or dusty hue; and the eyes are a pale
greyish-blue, with small pupils, and in very many
cases a dark mark round the iris.   The deep blue, any
pure blue, in fact, from forget-me-not to ultramarine,
is as rare in this commonest type as warm or bright
hair—chestnut, red, or gold; or as a brilliant skin.   The
skin is pallid, or dusky, or dirty-looking.   Even healthy
girls in their teens seldom have any colour, and the
exquisite roseate and carmine reds of other counties
are rare indeed.   The best-looking girls at the time of
life when they come nearest to being pretty, when
they are just growing into womanhood, have an un-
finished look which is almost pathetic.   One gets the
fancy that Nature had meant to make them nice-
looking, and finally becoming dissatisfied with her
work, left them to grow to maturity anyhow.   It is
pathetic, because there was little more to be done—a
rosier blush on the cheek, a touch of scarlet on the
lips, a little brightness and elasticity in the hair, a
pencil of sunlight to make the eyes sparkle.

In figure this woman is slim, too narrow across the
hips, too flat in the chest.   And she grows thinner with
years.   The number of lean, pale women of this type
in Hampshire is very remarkable.   You see them in
every village, women that appear almost fleshless, with
a parchment-like skin drawn tight over the bones of
the face, pale-blue, washed-out eyes, and thin, dead-
looking hair.   What is the reason of this leanness?   It

may be that the women of this blonde type are more
subject to poverty of blood than others; for the men,
though often thin, are not so excessively thin as
the women. Or it may be the effect of that kind of
poison which cottage women all over the country are
becoming increasingly fond of, and which is having so
deleterious an effect on the people in many counties—
the tea they drink. Poison it certainly is: two or
three cups a day of the black juice which they obtain
by boiling and brewing the coarse Indian teas at a
shilling a pound which they use, would kill me in less
than a week.

Or it may be partly the poison of tea and partly the
bad conditions, especially the want of proper food, in
the villages. One day on the downs near Winchester
I found a shepherd with his flock, a man of about fifty,
and as healthy and strong looking a fellow as I have
seen in Hampshire. Why was it, I asked him, that he
was the only man of his village I had seen with the
colour of red blood in his face? why did they look so
unwholesome generally? why were the women so thin,
and the children so stunted and colourless? He said
he didn't know, but thought that for one thing they did
not get enough to eat. "On the farm where I work,"
he said, "there are twelve of us—nine men, all married,
and three boys. My wages are thirteen shillings, with
a cottage and garden; I have no children, and I neither
drink nor smoke, and have not done so for eighteen
years. Yet I find the money is not too much. Of

the others, the eight married men all have children —one has got.six at home: they all smoke, and all make a practice of spending at least two evenings each week at the public-house." How, after paying for beer and tobacco, they could support their families on the few shillings that remained out of their wages was a puzzle to him.

But this is to digress. The prevalent blonde type I have tried to describe is best seen in the northern half of the county, but is not so accentuated on the east, north, and west borders as in the interior villages. If, as is commonly said, this people is Anglo-Saxon, it must at some early period have mixed its blood with that of a distinctly different race. This may have been the Belgic or Brythonic, but as shape and face are neither Celtic nor Saxon, the Brythons must have already been greatly modified by some older and different race which they, or the Goidels before them, had conquered and absorbed. It will be necessary to return to this point by-and-by.

Side by side with this, in a sense, dim and doubtful people, you find the unmistakable Saxon, the thick-set, heavy-looking, round-headed man with blue eyes and light hair and heavy drooping mustachios—a sort of terrestrial walrus who goes erect. He is not abun-dant as in Sussex, but is represented in almost any village, and in these villages he is always like a bull-dog or bull-terrier among hounds, lurchers, and many other varieties, including curs of low degree. Mentally, he is rather a dull dog, at all events deficient in the finer, more attractive qualities. Leaving aside the spiritual part, he is a good all-round man, tough and stubborn, one the naturalist may have no secret qualms about in treating as an animal. A being of strong animal nature, and too often in this brewer-ridden county a hard drinker. A very large proportion of the men in rural towns and villages with blotchy skins and watery or beery eyes are of this type. Even more offensive than the animality, the mindlessness, is that flicker of conscious superiority which lives in their expression. It is, I fancy, a survival of the old instinctive feeling of a conquering race amid the conquered.

Nature, we know, is everlastingly harking back, but here in Hampshire I cannot but think that this type, in spite of its very marked characters, is a very much muddied and degenerate form. One is led to this conclusion by occasionally meeting with an individual whose whole appearance is a revelation, and strikes the mind with a kind of astonishment, and one can only exclaim — there is nothing else to say—Here Nature has at length succeeded in reproducing the pure unadulterated form! Such a type I came upon one summer day on the high downs east of the Itchen.

He was a shepherd, a young fellow of twenty, about five feet eight in height, but looking short on account of his extraordinary breadth of shoulders and depth of chest. His arms were like a blacksmith's, and his legs thick, and his big head was round as a Dutch cheese.

He could, I imagined, have made a breach in the stone
wall near which I found him with his flock, if he had
lowered that hard round head and charged it like a
rhinoceros. His hair was light brown, and his face a
uniform rosy brown—in all Hampshire no man nor
woman had I seen so beautiful in colour, and his round,
keen, piercing eyes were of a wonderful blue—"eyes
like the sea." If this poor fellow, washed clean and
clothed becomingly in white flannels, had shown himself
in some great gathering at the Oval or some such place
on some great day, the common people would have
parted on either side to make way for him, and would
have regarded him with a kind of worship—an impulse
to kneel before him. There, on the downs, his appearance was almost grotesque in the dress he wore, made
of some fabric intended to last for ever, but now frayed,
worn to threads in places, and generally earth-coloured.
A small old cap, earth-coloured too, covered a portion
of his big, round head, and his ancient, lumpish, cracked
and clouted boots were like the hoofs of some extinct
large sort of horse which he had found fossilised among
the chalk hills. He had but eleven shillings a week,
and could not afford to spend much on dress. How he
could get enough to eat was a puzzle; he looked as if
he could devour half of one of his muttons at a meal,
washed down with a bucket of beer, without hurt to his
digestion.   In appearance he formed a startling contrast
to the people around him: they were in comparison a
worn-out, weary-looking race, dim-eyed, pale-faced, slow

in their movements, as if they had lost all joy and interest in life.

The sight of him taught me something I could not get from the books. The intensity of life in his eyes and whole expression; the rough-hewn face and rude, powerful form—rude but well balanced—the vigour in his every movement, enabled me to realise better than anything that history tells us what those men who came as strangers to these shores in the fifth century were really like, and how they could do what they did. They came, a few at a time, in open row-boats, with nothing but their rude weapons in their hands, and by pure muscular force, and because they were absolutely without fear and without compassion, and were mentally but little above a herd of buffaloes, they succeeded in conquering a great and populous country with centuries of civilisation behind it.

Talking with him, I was not surprised to find him a discontented man. He did not want to live in a town— he seemed not to know just what he wanted, or having but few words he did not know how to say it; but his mind was in a state of turmoil and revolt, and he could only curse the head shepherd, the bailiff, the farmer, and, to finish up, the lord of the manor. Probably he soon cast away his crook, and went off in search of some distant place, where he would be permitted to discharge the energy that seethed and bubbled in him—perhaps to bite the dust on the African veldt.

This, then, is one of the main facts to be noted in the

blonde Hampshire peasant—the great contrast between the small minority of persons of the Anglo-Saxon and of the prevalent type. It was long ago shown by Huxley that the English people generally are not Saxons in the shape of the head, and in all Saxon England the divergence has perhaps been greatest in this southern county. The oval-faced type, as I have said, is less pronounced as we approach the borders of Berkshire, and although the difference is not very great, it is quite perceptible; the Berkshire people are rather nearer to the common modified Saxon type of Oxfordshire and the Midlands generally.

In the southern half of Hampshire the dark-eyed, black-haired people are almost as common as the blonde, and in some localities they are actually in a majority. Visitors to the New Forest district often express astonishment at the darkness and "foreign" appearance of the people, and they sometimes form the mistaken idea that it is due to a strong element of gipsy blood. The darkest Hampshire peasant is always in shape of head and face the farthest removed from the gipsy type.

Among the dark people there are two distinct types, as there are two in the blonde, and it will be understood that I only mean two that are, in a measure, fixed and easily recognised types; for it must always be borne in mind that, outside of these distinctive forms, there is a heterogeneous crowd of persons of all shades and shapes of face and of great variety in features. These two dark types are—first, the small, narrow-headed person of brown skin, crow-black hair, and black eyes; of this rarest and most interesting type I shall speak last. Second, the person of average height, slightly oval face, and dark eyes and hair. The accompanying portrait of a young woman in a village on the Test is a good specimen of

A Hampshire Girl

this type. Now we find that this dark-haired, dark-eyed, and often dark-skinned people are in stature, figure, shape of head, and features exactly like the oval-faced blonde people already described. They are, light and dark, an intermediate type, and we can only say that they are one and the same people, the outcome of a long mixed race which has crystallised in this form unlike any of its originals; that the difference in colour is due to the fact that blue and black in the iris and black and brown in the hair very seldom mix, these colours being, as has been said, "mutually exclusive." They persist when everything else, down to the bony framework, has been modified and the original racial characters obliterated. Nevertheless, we see that these mutually exclusive colours do mix in some individuals both in the eyes and hair. In the grey-blue iris it appears as a very slight pigmentation, in most cases round the pupil, but in the hair it is more marked. Many, perhaps a majority, of the dark-eyed people we are now considering have some warm brown colour in their black hair; in members of the same family you will often find raven-black hair and brownish-black hair; and sometimes in three brothers or sisters you will find the two original colours, black and brown, and the intermediate very dark or brownish-black hair.

The brunette of this oval-faced type is also, as we have seen, deficient in colour, but, as a rule, she is more attractive than her light-eyed sister. This may be due to the appearance of a greater intensity of life in the dark eye; but it is also probable that there is almost always some difference in disposition, that black or dark pigment is correlated with a warmer, quicker, more sympathetic nature. The anthropologists tell us that very slight differences in intensity of pigmentation may correspond to relatively very great constitutional differences. One fact in reference to dark- and light-coloured people which I came upon in Hampshire, struck me as exceedingly curious, and has suggested the question: Is there in us, or in some of us, very deep down, and buried out of sight, but still occasionally coming to life and to the surface, an ancient feeling of repulsion or racial antipathy between black and blonde? Are there mental characteristics, too, that are "mutually exclusive"? Dark and light are mixed in very many of us, but, as Huxley has said, the constituents do not always rightly mix: as a rule, one side is strongest. With the dark side strongest in me, I search myself, and the only evidence I find of such a feeling is an ineradicable dislike of the shallow frosty blue eye: it makes me shiver, and seems to indicate a cold, petty, spiteful, and false nature. This may be merely a fancy or association, the colour resembling that of the frosty sky in winter. In many others the feeling appears to be more definite. I know blue-eyed persons of culture, liberal-minded, religious, charitable, lovers of all men, who declare that they cannot regard dark-eyed persons as being on the same level, morally, with the blue-eyed, and that they cannot dissociate black eyes from wickedness. This, too, may be fancy or association. But here in Hampshire I have been startled at some things I have heard spoken by dark-eyed people about blondes. Not of the mitigated Hampshire blonde, with that dimness in the colour of his skin, and eyes, and hair, but of the more vivid type with brighter blue eyes, and brighter or more fiery hair, and the light skin to match. What I have heard was to this effect:—

"Perhaps it will be all right in the end—we hope it will: he says he will marry her and give her a home. But you never know where you are with a man of that colour—I'll believe it when I see it."

"Yes, he seems all right, and speaks well, and promises to pay me the money. But look at the colour of his eyes! No, I can't trust him."

"He's a very nice person, I have no doubt, but his eyes and hair are enough for me," &c., &c.

Even this may be merely the effect of that enmity or suspicion with which the stranger, or "foreigner," as he is called, is often regarded in rural districts. The person from another county, or from a distance, unrelated to any one in the community, is always a foreigner, and the foreign taint may descend to the children: may it not be that in Hampshire any one with bright colour in eyes, hair, and skin is also by association regarded as a foreigner?

It remains to speak of the last of the four distinct types, the least common and most interesting of all— the small, narrow-headed man with very black hair, black eyes, and brown skin.

We are deeply indebted to the anthropologists who have, so to speak, torn up the books of History, and are re-telling the story of Man on earth: we admire them for their patient industry, and because they have gone bravely on with their self-appointed task, one peculiarly

R

difficult in this land of many mixed races, heedless of the scoffs of the learned or of those who derive their learning from books alone, and mock at men whose documents are "bones and skins." But we sometimes see that they (the anthropologists) have not yet wholly emancipated themselves from the old written false-hoods when they tell us, as they frequently do, that the Iberian in this country survives only in the west and the north. They refer to the small, swarthy Welshman; to the so-called "black Celt" in Ireland, west of the Shannon; to the small black Yorkshireman of the Dales, and to the small black Highlander; and the explanation is that in these localities remnants of the dark men of the Iberian race who inhabited Britain in the Neolithic period, were never absorbed by the conquerors; that, in fact, like the small existing herds of indigenous white cattle, they have preserved their peculiar physical char-acter down to the present time by remaining unmixed with the surrounding blue-eyed people. But this type is not confined to these isolated spots in the west and north; it is found here, there, and everywhere, especially in the southern counties of England: you cannot go about among the peasants of Hampshire, Wiltshire, and Dorset without meeting examples of it, and here, at all events, it cannot be said that the ancient British people were not absorbed. They, the remnant that escaped extermination, were absorbed by the blue-eyed, broad-headed, tall men, the Goidels we suppose, who occupied the country at the beginning of the Bronze Age; and

the absorbers were in their turn absorbed by another
blue-eyed race; and these by still another or by others.
The only explanation appears to be that this type is
persistent beyond all others, and that a very little black
blood, after being mixed and re-mixed with blonde for
centuries, even for hundreds of generations, may, whenever the right conditions occur, reproduce the vanished
type in its original form.

Time brings about its revenges in many strange ways:
we see that there is a continuous and an increasing
migration from Wales and the Highlands into all the
big towns in England, and this large and growing Celtic
element will undoubtedly have a great effect on the
population in time, making it less Saxon and more
Celtic than it has been these thousand years past and
upwards. But in all the people, Celtic, Anglo-Saxon,
Dane, or what not, there is that older constituent—
infinitely older and perhaps infinitely more persistent;
and this too, albeit in a subtler way, may be working
in us to recover its long-lost world. That it has gone
far in this direction in Spain, where the blue eye is
threatened with extinction, and in the greater portion
if not all, of France, there appears to be some evidence
to show. Here, where the Neolithic people were more
nearly exterminated and the remnant more completely
absorbed, the return may be very much slower. But
when we find, as we do in Hampshire and many other
counties, that this constituent in the blood of the
people, after mixture for untold ages with so many

other bloods of so many conquering races, has not only been potent to modify the entire population, but is able to reproduce the old type in its pristine purity; and when we almost invariably find that these ancients born again are better men than those in whom other racial characters predominate—more intelligent, versatile, adaptive, temperate, and usually tougher and longer lived, it becomes possible to believe that in the remote future—there are thousands of years for this little black leaven to work—these islands will once more be inhabited by a race of men of the Neolithic type.

In speaking of the character, physical and mental, of the men of distinctly Iberian type, I must confess that I write only from my own observation, and that I am hardly justified in founding general statements on an acquaintance with a very limited number of persons. My experience is that the men of this type have, generally speaking, more character than their neighbours, and are certainly very much more interesting. In recalling individuals of the peasant class who have most attracted me, with whom I have become intimate and in some instances formed lasting friendships, I find that of twenty-five to thirty no fewer than nine are of this type. Of this number four are natives of Hampshire, while the other five, oddly enough, belong to five different counties. But I do not judge only from these few individuals: a rambler about the country who seldom stays many days in one village or spot cannot become intimately acquainted with the cottagers. I judge partly from the few I know well, and partly from a very much larger number of individuals I have met casually or have known slightly. What I am certain of is that the men of this type, as a rule, differ mentally as widely as they do physically from persons of other commoner types. The Iberian, as I know him in southern and south-western England, is, as I have said, more intelligent, or at all events, quicker; his brains are nimbler although perhaps not so retentive or so practical as the slower Saxon's. Apart from that point, he has more imagination, detachment, sympathy—the qualities which attract and make you glad to know a man and to form a friendship with him in whatever class he may be. Why is it, one is sometimes asked, that one can often know and talk with a Spaniard or Frenchman without any feeling of class distinction, any consciousness of a barrier, although the man may be nothing but a workman, while with English peasants this freedom and ease between man and man is impossible? It *is* possible in the case of the man we are considering simply because of those qualities I have named, which he shares with those of his own race on the continent.

I have found that when one member of a family of mixed light and dark blood is of the distinctly Iberian type, this one will almost invariably take a peculiar and in some ways a superior position in the circle. The woman especially exhibits a liveliness, humour,

and variety rare indeed among persons born in the
peasant class. She entertains the visitor, or takes the
leading part, and her slow-witted sisters regard her
with a kind of puzzled admiration. They are sisters,
yet unrelated: their very blood differs in specific
gravity, and their bodily differences correspond to a
mental and spiritual unlikeness. In my intercourse
with people in the southern counties I have sometimes
been reminded of Huxley and his account of his parents
contained in a private letter to Havelock Ellis. His
father, he said, was a fresh-coloured, grey-eyed War-
wickshire man. " My mother came of Wiltshire people.
Except for being somewhat taller than the average type,
she was a typical example of the Iberian variety—dark,
thin, rapid in all her ways, and with the most piercing
black eyes I have seen in anybody's head. Mentally
and physically (except in the matter of the beautiful
eyes) I am a piece of my mother, and except for my
stature . . . I should do very well for a 'black Celt'—
supposed to be the worst variety of that type."

The contrast between persons of this type and Saxon
or blonde has often seemed to me greatest in childhood,
since the blonde at that period, even in Hampshire,
is apt to be a delicate pink and white, whereas the
individual of strongly-marked Iberian character is very
dark from birth. I will, to conclude this perhaps im-
prudent chapter, give an instance in point.

Walking one day through the small rustic village
of Martyr Worthy, near Winchester, I saw a little

girl of nine or ten sitting on the grass at the side of the wide green roadway in the middle of the village, engaged in binding flowers round her hat. She was slim, and had a thin oval face, dark in colour as any dark Spanish child, or any French child in the "black provinces"; and she had, too, the soft melancholy

black eye which is the chief beauty of the Spanish, and her loose hair was intensely black. Even here where dark eyes and dark hair are so common, her darkness was wonderful by contrast with a second little girl of round, chubby, rosy face, pale-yellowish hair, and wide-open blue surprised eyes, who stood by her side watching her at her task. The flowers were lying in a heap at her side; she had wound a long slender spray of traveller's joy round her brown straw hat, and was now weaving in lychnis and veronica, with other small red and blue blossoms, to improve her garland. I found to my surprise on questioning her that she knew the names of the flowers she had collected. An English village child, but in that Spanish darkness and beauty, and in her grace and her pretty occupation, how very un-English she seemed!

CHAPTER XII

THERE are no more refreshing places in Hampshire,
one might almost say in England, than the green level
valleys of the Test and Itchen that wind, alternately
widening and narrowing, through the downland country
to Southampton Water. Twin rivers they may be
called, flowing at no great distance apart through the
same kind of country, and closely alike in their
general features: land and water intermixed—greenest
water-meadows and crystal currents that divide and
subdivide and join again, and again separate, forming
many a miniature island and long slip of wet meadow
with streams on either side. At all times refreshing
to the sight and pleasant to dwell by, they are best

When it is summer and the green is deep.

Greens of darkest bulrushes, tipped with bright brown
panicles, growing in masses where the water is wide
and shallowest; of grey-green graceful reeds, and of

tallest reed-mace with dark velvety brown spikes; behind them all, bushes and trees—silvery-leafed willow
and poplar, and dark alder, and old thorns and brambles
in tangled masses; and always in the foreground
lighter and brighter sedges, glaucous green flags,
mixed with great hemp agrimony, with flesh-coloured,

white-powdered flowers, and big-leafed comfrey, and
scores of other water and moisture-loving plants.

Through this vegetation, this infinite variety of refreshing greens and graceful forms, flow the rapid rivers,
crystal clear and cold from the white chalk, a most
beautiful water, with floating water-grass in it — the
foscinating Poa fluviatilis which, rooted in the pebbly
bed, looks like green loosened wind-blown hair swaying
and trembling in the ever-crinkled, swift current.

They are not long rivers—the Test and Itchen—but long enough for men with unfevered blood in their veins to find sweet and peaceful homes on their margins. I think I know quite a dozen villages on the former stream, and fifteen or sixteen on the latter, in any one of which I could spend long years in perfect contentment. There are towns, too, ancient Romsey and Winchester, and modern hideous Eastleigh; but the little centres are best to live in. These are, indeed, among the most characteristic Hampshire villages; mostly small, with old thatched cottages, unlike, yet harmonising, irregularly placed along the roadside; each with its lowly walls set among gaily coloured flowers; the farm with its rural sounds and smells, its big horses and milch-cows led and driven along the quiet streets; the small ancient church with its low, square tower, or grey shingled spire; and great trees standing singly or in groups or rows—oak and elm and ash; and often some ivy-grown relic of antiquity—ivy, indeed, everywhere. The charm of these villages that look as natural and one with the scene as chalk down and trees and green meadows, and have an air of immemorial quiet and a human life that is part of nature's life, unstrenuous, slow and sweet, has not yet been greatly disturbed. It is not here as in some parts of Hampshire, and as it is pretty well everywhere in Surrey, that most favoured county, the Xanadu of the mighty ones of the money-market, where they oftenest decree their lordly pleasure-domes. Those vast red-brick habitations of the Kubla

Khans of the city which stare and glare at you from all openings in pine woods, across wide heaths and commons, and from hill-sides and hill-tops, produce the idea that they were turned out complete at some stupendous manufactory of houses at a distance, and sent out by the hundred to be set up wherever wanted, and where they are almost always utterly out of keeping with their surroundings, and consequently a blot on and a disfigurement of the landscape.

Happily the downland slopes overlooking these green valleys have so far been neglected by the class of persons who live in mansions; for. the time being they are ours, and by "ours" I mean all those who love and reverence this earth. But which of the two is best I cannot say. One prefers the Test and another the Itchen, doubtless because in a matter of this kind the earth-lover will invariably prefer the spot he knows most intimately; and for this reason, much as I love the Test, long as I would linger by it, I love the Itchen more, having had a closer intimacy with it. I dare say that some of my friends, old Wykehamists, who as boys caught their first trout close by the ancient sacred city and have kept up their acquaintance with its crystal currents, will laugh at me for writing as I do. But there are places, as there are faces, which draw the soul, and with which, in a little while, one becomes strangely intimate.

The first English cathedral I ever saw was that of Winchester: that was a long time ago; it was then

and on a few subsequent occasions that I had glimpses
of the river that runs by it. They were like momentary
sights of a beautiful face, caught in passing, of some
person unknown. Then it happened that in June 1900,
cycling Londonwards from Beaulieu and the coast by
Lymington, I came to the valley, and to a village about
half-way between Winchester and Alresford, on a visit
to friends in their summer fishing retreat.

They had told me about their cottage, which serves
them all the best purposes of a lodge in the vast wilderness. Fortunately in this case the "boundless contiguity
of shade" of the woods is some little distance away, on
the other side of the ever green Itchen valley, which,
narrowing at this spot, is not much more than a couple
of hundred yards wide. A long field's length away
from the cottage is the little ancient, rustic, tree-hidden
village. The cottage, too, is pretty well hidden by trees,
and has the reed and sedge and grass green valley and
swift river before it, and behind and on each side green
fields and old untrimmed hedges with a few old oak
trees growing both in the hedgerows and the fields.
There is also an ancient avenue of limes which leads
nowhere and whose origin is forgotten. The ground
under the trees is overgrown with long grass and
nettles and burdock; nobody comes or goes by it, it is
only used by the cattle, the white and roan and strawberry shorthorns that graze in the fields and stand in
the shade of the limes on very hot days. Nor is there
any way or path to the cottage; but one must go and

come over the green fields, wet or dry. The avenue ends just at the point where the gently sloping chalk down touches the level valley, and the half-hidden, low-roofed cottage stands just there, with the shadow of the last two lime trees falling on it at one side. It was an ideal spot for a nature-lover and an angler to pitch his tent upon. Here a small plot of ground, including the end of the lime-tree avenue, was marked out, a hedge of sweetbriar planted round it, the cottage erected, and a green lawn made before it on the river side, and beds of roses planted at the back.

Nothing more—no gravel walks; no startling scarlet geraniums, no lobelias, no cinerarias, no calceolarias, nor other gardeners' abominations to hurt one's eyes and make one's head ache. And no dog, nor cat, nor chick, nor child—only the wild birds to keep one company. They knew how to appreciate its shelter and solitariness; they were all about it, and built their nests amid the great green masses of ivy, honeysuckle, Virginia creeper, rose, and wild clematis which covered the trellised walls and part of the red roof with a twelve years' luxuriant growth.

To this delectable spot I returned on July 21 to see the changeful summer of 1900 out, my friends having gone north and left me their cottage for a habitation.

"There is the wind on the heath, brother," and one heartily agrees with the half-mythical Petulengro that it is a very good thing; it had, indeed, been blowing off and in my face for many months past; and from shadeless heaths and windy downs, and last of all, from the intolerable heat and dusty desolation of London in mid-July, it was a delightful change to this valley.

During the very hot days that followed it was pleasure enough to sit in the shade of the limes most of the day; there was coolness, silence, melody, fragrance; and, always before me, the sight of that moist green valley, which made one cool simply to look at it, and never wholly lost its novelty. The grass and herbage grow so luxuriantly in the water-meadows that the cows grazing there were half-hidden in their depth; and the green was tinged with the purple of seeding grasses, and red of dock and sorrel, and was everywhere splashed with creamy white of meadow-sweet. The channels of the swift many-channelled river were fringed with the livelier green of sedges and reed-mace, and darkest green of bulrushes, and restful grey of reeds not yet in flower.

The old limes were now in their fullest bloom; and the hotter the day the greater the fragrance, the flower, unlike the woodbine and sweetbriar, needing no dew nor rain to bring out its deliciousness. To me, sitting there, it was at the same time a bath and atmosphere of sweetness, but it was very much more than that to all the honey-eating insects in the neighbourhood. Their murmur was loud all day till dark, and from the lower branches that touched the grass

with leaf and flower to their very tops the trees were
peopled with tens and with hundreds of thousands of
bees. Where they all came from was a mystery;
somewhere there should be a great harvest of honey
and wax as a result of all this noise and activity. It
was a soothing noise, according with an idle man's
mood in the July weather; and it harmonised with,
forming, so to speak, an appropriate background to,
the various distinct and individual sounds of bird
life.

The birds were many, and the tree under which I sat
was their favourite resting-place; for not only was it the
largest of the limes, but it was the last of the row, and
overlooked the valley, so that when they flew across
from the wood on the other side they mostly came to
it. It was a very noble tree, eighteen feet in circum-
ference near the ground; at about twenty feet from the
root, the trunk divided into two central boles and
several of lesser size, and these all threw out long horizontal and drooping branches, the lowest of which
feathered down to the grass. One sat as in a vast
pavilion, and looked up to a height of sixty or seventy
feet through wide spaces of shadow and green sunlight,
and sunlit golden-green foliage and honey-coloured
blossom, contrasting with brown branches and with
masses of darkest mistletoe.

Among the constant succession of bird visitors to the
tree above me were the three pigeons—ring-dove, stock-
dove, and turtle-dove; finches, tree-warblers, tits of

four species, and the wren, tree-creeper, nut-hatch, and many more. The best vocalists had ceased singing; the last nightingale I had heard utter its full song was in the oak woods of Beaulieu on June 27: and now all the tree-warblers, and with them chaffinch, thrush, blackbird, and robin, had become silent. The wren was the leading songster, beginning his bright music at four o'clock in the morning, and the others, still in song, that visited me were the greenfinch, goldfinch, swallow, dunnock, and cirl bunting. From my seat I could also hear the songs in the valley of the reed and sedge warblers, reed-bunting, and grasshopper-warbler. These, and the polyglot starling, and cooing and crooning doves, made the last days of July at this spot seem not the silent season we are accustomed to call it.

Of these singers the goldfinch was the most pleasing. The bird that sang near me had assisted in rearing a brood in a nest on a low branch a few yards away, but he still returned from the fields at intervals to sing; and seen, as I now saw him a dozen times a day, perched among the lime leaves and blossoms at the end of a slender bough, in his black and gold and crimson livery, he was by far the prettiest of my feathered visitors.

But the cirl bunting, the inferior singer, interested me most, for I am somewhat partial to the buntings, and he is the best of them, and the one I knew least about from personal observation.

On my way hither at the end of June, somewhere

s

between Romsey and Winchester, a cock cirl bunting in fine plumage flew up before me and perched on the wire of a roadside fence. It was a welcome encounter, and, alighting, I stood for some time watching him. I did not know that I was in a district where this pretty species is more numerous than in any other place in England—as common, in fact, as the universal yellow-hammer, and commoner than the more local corn bunting. Here in July and August, in the course of an afternoon's walk, in any place where there are trees and grass fields, one can count on hearing half-a-dozen birds sing, every one of them probably the parent of a nest full of young. For this is the cirl bunting's pleasant habit. He assists in feeding and safeguarding the young, even as other songsters do who cease singing when this burden is laid upon them; but he is a bird of placid disposition, and takes his task more quietly than most; and, after returning from the fields with several grasshoppers in his throat and beak and feeding his fledglings, he takes a rest, and at intervals in the day flies to his favourite tree, and repeats his blithe little song half-a-dozen times.

The song is not quite accurately described in the standard ornithological works as exactly like that of the yellow-hammer, only without the thin, drawn-out note at the end, and therefore inferior—the little bit of bread, but without the cheese. It certainly resembles the yellow-hammer's song, being a short note, a musical chirp, rapidly repeated several times. But the yellow-hammer

varies his song as to its time, the notes being sometimes
fast and sometimes slow. The cirl's song is always the
same in this respect, and is always a more rapid song

CIRL BUNTING

than that of the other species. So rapid is it that,
heard at a distance, it acquires almost the character of
a long trill. In quality, too, it is the better song—
clearer, brighter, brisker—and it carries farther; on

still mornings I could hear one bird's song very distinctly at a distance of two hundred and fifty yards. The only good description of the cirl bunting's song—as well as the best general account of the bird's habits—which I have found, is in J. C. Bellamy's *Natural History of South Devon* (Plymouth and London), 1839, probably a forgotten book.

The best singer among the British buntings, he is also to my mind the prettiest bird. When he is described as black and brown, and lemon and sulphur - yellow, and olive and lavender-grey, and chestnut-red, we are apt to think that the effect of so many colours thrown upon his small body cannot be very pleasing. But it is not so; these various colours are so harmoniously disposed, and have, in the lighter and brighter hues in the living bird, such a flower-like freshness and delicacy, that the effect is really charming.

When, in June, I first visited the cottage, my host took me into his dressing-room, and from it we watched a pair of cirl buntings bring food to their young in a nest in a small cypress standing just five yards from the window. The young birds were in the pinfeather stage, but they were unfortunately taken a very few days later by a rat, or stoat, or by that winged nest-robber the jackdaw, whose small cunning grey eyes are able to see into so many hidden things.

The birds themselves did not grieve overlong at their loss: the day after the nest was robbed the cock was heard singing—and he continued to sing every day from his favourite tree, an old black poplar growing outside the sweetbriar hedge in front of the cottage.

About this bird of a brave and cheerful disposition, more will have to be said in the next chapter. It is, or was, my desire to describe events in the valley at this changeful period from late July to October in the order of their occurrence, but in all the rest of the present chapter, which will be given to the river birds exclusively, the order must be broken.

Undoubtedly the three commonest water birds inhabiting inland waters throughout England are the coot, moor-hen, and dabchick, or little grebe; and on account of their abundance and general distribution they are almost as familiar as our domestic birds. Yet one never grows tired of seeing and hearing them, as we do of noting the actions of other species that inhabit the same places; and the reason for this—a very odd reason it seems!—is because these three common birds, members of two orders which the modern scientific zoologist has set down among the lowest, and therefore, as he tells us, most stupid, of the feathered inhabitants of the globe, do actually exhibit a quicker intelligence and greater variety in their actions and habits than the species which are accounted their superiors.

The coot is not so abundant as the other two; also he is less varied in his colour, and less lively in his motions, and consequently attracts us less. The moorhen is the most engaging, as well as the commonest—a bird concerning which more entertaining matter has

been related in our Natural Histories than of any other
native species. And I now saw a great deal of him,
and of the other two as well. From the cottage
windows, and from the lawn outside, one looked upon
the main current of the river, and there were the birds
always in sight; and when not looking one could hear
them. Without paying particular attention to them
their presence in the river was a constant source of
interest and amusement.

At one spot, where the stream made a slight bend,
the floating water-weeds brought down by the current
were always being caught by scattered bulrushes growing a few feet from the edge, the arrested weeds formed
a minute group of islets, and on these convenient little
refuges and resting-places in the waterway, a dozen or
more of the birds could be seen at most times. The
old coots would stand on the floating weeds and preen
and preen their plumage by the hour. They were like
mermaids, for ever combing out their locks, and had
the clear stream for a mirror. The dull-brown, white-
breasted young coots, now fully grown, would meanwhile swim about picking up their own food. The
moor-hens were with them, preening and feeding, and
one had its nest there. It was a very big conspicuous
nest, built up on a bunch of floating weeds, and formed,
when the bird was sitting on its eggs, a pretty and
curious object; for every day fresh bright-green sedge
leaves were plucked and woven round it, and on that
high bright-green nest, as on a throne, the bird sat, and

when I went near the edge of the water, she (or he) would flirt her tail to display the snowy-white under feathers, and nod her head, and stand up as if to display her pretty green legs, so as to let me see and admire all her colours; and finally, not being at all shy, she would settle quietly down again.

The little grebes, too, had chosen that spot to build on. Poor little grebes! how they worked and sat, and built and sat again, all the summer long. And all along the river it was the same thing — the grebes industriously making their nests, and trying ever so hard to hatch their eggs; and then at intervals of a few days the ruthless water-keeper would come by with his long fatal pole to dash their hopes. For whenever he saw a suspicious-looking bunch of dead floating weeds which might be a grebe's nest, down would come the end of the pole on it, and the eggs would be spilt out of the wet bed, and rolled down by the swift water to the sea. ' And then the birds would cheerfully set to work again at the very same spot: but it was never easy to tell which bunch of wet weeds their eggs were hidden in. Watching with a glass I could see the hen on her eggs, but if any person approached she would hastily pull the wet weeds from the edge over them, and slip into the water, diving and going away to some distance. While the female sat the male was always busy, diving and catching little fishes; he would dive down in one spot, and suddenly pop up a couple of yards away, right among the coots and moor-hens. This Jack-in-the-box action on his part never upset their nerves. They took not the slightest notice of him, and were altogether a more or less happy family, all very tolerant of each other's little eccentricities.

The little grebe fished for himself and for his sitting mate; he never seemed so happy and proud as when he was swimming to her, patiently sitting on her wet nest, with a little silvery fish in his beak. He also fished for old decaying weeds, which he fetched up from the bottom to add to the nest. Whenever he popped up among or near the other birds with an old rag of a weed in his beak, one or two of the grown-up young coots would try to take it from him; and seeing them gaining on him he would dive down to come up in another place, still clinging to the old rag half a yard long; and again the chase would be renewed, and again he would dive; until at last, after many narrow escapes and much strategy, the nest would be gained, and the sitting bird would take the weed from him and draw it up and tuck it round her, pleased with his devotedness, and at the sight of his triumph over the coots. As a rule, after giving her something—a little fish, or a wet weed to pull up and make herself comfortable with—they would join their voices in that long trilling cry of theirs, like a metallic, musical-sounding policeman's rattle.

It was not in a mere frolicsome spirit that the young coots hunted the dabchick with his weed, but

A more or less Happy Family

rather, as I imagine, because the white succulent stems of aquatic plants growing deep in the water is their favourite food; they are accustomed to have it dived for by their parents and brought up to them, and they never appear to get enough to satisfy them; but when they are big, and their parents refuse to slave for them, they seem to want to make the little grebes their fishers for succulent stems.

One day in August 1899, I witnessed a pretty little bird comedy at the Pen Ponds, in Richmond Park, which seemed to throw a strong light on the inner or domestic life of the coot. For a space of twenty minutes I watched an old coot industriously diving and bringing up the white parts of the stems of Polygonum persicaria, which grows abundantly there, together with the rarer more beautiful Lymnanthemum nymphoides, which is called Lymnanth for short. I prefer an English name for a British plant, an exceedingly attractive one in this case, and so beg leave to call it Water-crocus. The old bird was attended by a full-grown young one, which she was feeding, and the unfailing diligence and quickness of the parent were as wonderful to see as the gluttonous disposition of its offspring. The old coot dived at least three times every minute, and each time came up with a clean white stem, the thickness of a stout clay pipe-stem, cut the proper length—about three to four inches. This the young bird would take and instantly swallow; but before it was well down his throat the old bird would be gone for another. I was with a friend, and we wondered when its devouring cormorant appetite would be appeased, and how its maw could contain so much food; we also compared it to a hungry Italian greedily sucking down macaroni.

While this was going on a second young bird had been on the old nest on the little island in the lake, quietly dozing; and at length this one got off his dozing-place, and swam out to where the weed-fishing and feeding were in progress. As he came up, the old coot rose with a white stem in her beak, which the new-comer pushed forward to take; but the other thrust himself before him, and, snatching the stem from his parent's beak, swallowed it himself. The old coot remained perfectly motionless for a space of about four seconds, looking fixedly at the greedy one who had been gorging for twenty minutes yet refused to give place to the other. Then very suddenly, and with incredible fury, she dashed at and began hunting him over the pond. In vain he rose up and flew over the water, beating the surface with his feet, uttering cries of terror; in vain he dived; again and again she overtook and dealt him the most savage blows with her sharp beak, until, her anger thoroughly appeased and the punishment completed, she swam back to the second bird, waiting quietly at the same spot for her return, and began once more diving for white stems of the Polygonum.

Never again, we said, would the greedy young bird behave in the unmannerly way which had brought so terrible a castigation upon him! The coot is certainly a good mother who does not spoil her child by sparing the rod. And this is the bird which our comparative anatomists, after pulling it to pieces, tell us is a small-brained, unintelligent creature; and which old Michael Drayton, who, being a poet, ought to have known better, described as " a formal brainless ass " !

To come back to the Itchen birds. The little group, or happy family, I have described was but one of the many groups of the same kind existing all along the river; and these separate groups, though at a distance from each other, and not exactly on visiting terms, each being jealous of its own stretch of water, yet kept up a sort of neighbourly intercourse in their own way. Single cries were heard at all times from different points; but once or two or three times in the day a cry of a coot or a moor-hen would be responded to by a bird at a distance; then another would take it up at a more distant point, and another still, until cries answering cries would be heard all along the stream. At such times the voice of the skulking water-rail would be audible too, but whether this excessively secretive bird had any social relations with the others beyond joining in the general greeting and outcry I could not discover. Thus, all these separate little groups, composed of three different species, were like the members of one tribe or people broken up into

families; and altogether it seemed that their lines had
fallen to them in pleasant places, although it cannot
be said that the placid current of their existence was
never troubled.

I know not what happened to disturb them, but
sometimes all at once cries were heard which were
unmistakably emitted in anger, and sounds of splashing and struggling among the sedges and bulrushes;
and the rushes would be swayed about this way and
that, and birds would appear in hot pursuit of one
another over the water; and then, just when one was
in the midst of wondering what all this fury in their
cooty breasts could be about, lo! it would all be over,
and the little grebe would be busy catching his silvery
fishes; and the moor-hen, pleased as ever at her
own prettiness, nodding and prinking and flirting her
feathers; and the coot, as usual, mermaid-like, combing
out her slate-coloured tresses.

We have seen that of these three species the little
grebe was not so happy as the others, owing to his
taste for little fishes being offensive to the fish-breeder
and preserver. When I first saw how this river was
watched over by the water-keepers, I came to the
conclusion that very few or no dabchicks would succeed in hatching any young. And none were hatched
until August, and then to my surprise I heard at
one point the small, plaintive *peep-peep* of the young
birds crying to be fed. One little grebe, more cunning or more fortunate than the others, had at last

succeeded in bringing off her young; and once out
of their shells they were safe.   But by-and-by the
little duckling-like sound was heard at another point,
and then at another; and this continued in September, until, by the middle of that month, you
could walk miles along the river, and before you left
the sound of one little brood hungrily crying to be
fed behind you, the little *peep-peep* of another brood
would begin to be heard in advance of you.

Often enough it is "dogged as does it" in bird as
well as in human affairs, and never had birds more
deserved to succeed than these dogged little grebes.
I doubt if a single pair failed to bring out at least
a couple of young by the end of September.   And
at that date you could see young birds apparently
just out of the shell, while those that had been
hatched in August were full grown.

About the habits of the little grebe, as about those
of the moor-hen, many curious and entertaining things
have been written; but what amused me most in these
birds, when I watched them in late September on
the Itchen, was the skilful way in which the parent
bird taught her grown-up young ones to fish.   At
an early period the fishes given to the downy young
are very small, and are always well bruised in the
beak before the young bird is allowed to take it,
however eager he may be to seize it.   Afterwards,
when the young are more grown, the size of the
fishes is increased, and they are less and less bruised,

although always killed. Finally, the young has to
be taught to catch for himself; and at first he does
not appear to have any aptitude for such a task, or
any desire to acquire it. He is tormented with
hunger, and all he knows is that his parent can catch
fish for him, and his only desire is that she shall
go on catching them as fast as he can swallow them.
And she catches him a fish, and gives it to him,
but, oh·mockery! it was not really dead this time,
and instantly falls into the water and is lost! Not
hopelessly lost, however, for down she goes like lightning, and comes up in ten seconds with it again. And
he takes and drops it again, and looks stupid, and
again she recovers and gives it to him. How many
hundreds of times, I wonder, must this lesson be repeated before the young grebe finds out how to keep
and to kill? Yet that is after all only the beginning of his education. The main thing is that he
must be taught to dive after the fishes he lets fall,
and he appears to have no inclination, no intuitive
impulse, to do such a thing. A small, quite dead
fish must be given him carelessly, so that it shall
fall, and he must be taught to pick up a fallen
morsel from the surface; but from that first simple
act to the swift plunge and long chase after and
capture of uninjured vigorous fishes, what an immense distance there is! It is, however, probable
that, after the first reluctance of the young bird
has been overcome, and a habit of diving after

escaped fishes acquired, he makes exceedingly rapid progress

But, even after the completion of his education, when he is independent of his parents, and quick and sure as they at capturing fishes down in their own dim element, is it not still a puzzle and a mystery that such a thing can be done? And here I speak not only of the little grebe, but of all birds that dive after fishes, and pursue and capture them in fresh or salt water. We see how a kingfisher takes his prey, or a tern, or gannet, or osprey, by dropping upon it when it swims near the surface; he takes his fish by surprise, as a sparrow-hawk takes the birds he preys upon. But no specialisation can make an air-breathing, feathered bird an equal of the fish under water. One can see at a glance in any clear stream that any fish can out-distance any bird, darting off with the least effort so swiftly as almost to elude the sight, while the fastest bird under water moves but little faster than a water rat.

The explanation, I believe, is that the paralysing effect on many small, persecuted creatures in the presence of, or when pursued by, their natural enemies and devourers, is as common under as above water. I have distinctly seen this when watching fish-eating birds being fed at the Zoological Gardens in glass tanks. The appearance of the bird when he dives strikes an instant terror into them; and it may then be seen that those which endeavour to escape are no longer in possession of their full powers, and their efforts to fly from the enemy are like those of the mouse and vole when a weasel is on their track, or of a frog when pursued by a snake ; while others remain suspended in the water, quite motionless, until seized and swallowed.

Bransbury Mill.

CHAPTER XIII

Morning in the valley—Abundance of swifts—Unlikeness to other birds—Mayfly and swallows—Mayfly and swift—Bad weather and hail—Swallows in the rain—Sand martins—An orphaned blackbird—Tamed by feeding—Survival of gregarious instinct in young blackbirds—Blackbird's good-night—Cirl buntings—Breeding habits and language—Habits of the young—Reed bunting—Beautiful weather—The oak in August.

DURING the month of July the swift was the most abundant and most constantly before us of all our Itchen valley birds. In the morning he was not there. We had the pigeons then, all three species—ring-dove, stock-dove and turtle-dove — being abundant in the woods on the opposite side of the valley, and from four o'clock to six was the time of their morning concert, when the still air was filled with the human-like musical sound of their multitudinous voices mingled in one voice. An hour or two later, as the air grew warmer, the swifts would begin to arrive to fly up and down the stream incessantly until dark, feasting on the gnats and ephemeræ that swarmed over the water during those hot days of late summer. Doubtless these birds come every day from all the towns, villages, and farm-houses scattered over a very broad strip of country on either side of the Itchen. Never had I seen swifts so numerous; looking down on the valley from any point one had hundreds of birds in sight at once, all swiftly flying up and down stream, but when the sight was kept fixed on any one bird, it could be seen that he went but a short distance—fifty to a hundred yards—then turned back. Thus each bird had a very limited range, and probably each returned to his accustomed place or beat every day.

These swifts are very much in the angler's way. Frequently they get entangled in the line and are brought down, but are seldom injured. During one day's fishing my friend here had three swifts to disengage from his line. On releasing one of these birds he watched its movements, and saw it fly up stream a distance of about forty yards, then double back, mechanically going on with its fly-hunting up and down stream just as if nothing had happened.

It may be said of swifts, as Bates said of humming-birds, that, mentally, they are more like bees than birds. The infallible, unchangeable way in which they, machine-like, perform all their actions, and their absolute unteachableness, are certainly insect-like. They are indeed so highly specialised and perfected in their own line; and, on account of their marvellous powers of flight, so removed from all friction in that atmosphere in which they live and move, above the complex and wit-sharpening conditions in which the more terrestrial creatures of their class exist, as to be practically independent of experience.

It is known that for some time the mayfly has been

decreasing, and in places disappearing altogether from these Hampshire streams, and it is believed and said by some of those who are concerned at these changes that the swallow is accountable for them. I do not know whether they have invented this brilliant idea themselves or have taken it ready-made from the water-keeper. Probably the last, since he, the water-keeper, is apt to regard all creatures that come to the waters where his sacred fishes are with a dull, hostile suspicion though in some cases he is not above adding to his income by taking a few trout himself—not indeed with the dry fly, which is useless at night, but with the shoe-net. In any case the question of exterminating the swallows in all the villages near the rivers has been seriously considered. Now, it is rather odd that this notion about the swallow—the martin is of course included—should have got about just when this bird has itself fallen on evil times and is decreasing with us. This decrease has, in all the parts of the country best known to me, become increasingly rapid during the last few years, and is probably due to new and improved methods of taking the birds wholesale during migration in France and Spain. Putting that matter aside, I should like to ask those gentlemen who have decreed, or would like to decree, the abolition of the swallow in all the riverside villages, what they propose to do about the swift?

One day last June (1902) I was walking with two friends by the Itchen, when a little below the village of

Ovington we sat down to rest and to enjoy a gleam of sunshine which happened to visit the world about noon that day. We sat down on a little wooden bridge over the main current and fell to watching the swifts, which were abundant, flying up and down just over our heads and, swift-like, paying no more heed to us than if we

had been three wooden posts or three cows. We noticed that ephemeræ of three or four species were rising up, and, borne by a light wind, drifting down-stream towards us and past us; and after watching these flies for some time we found that not one of them escaped. Small and grey, or dun, or water-coloured and well-nigh invisible, or large and yellow and con-spicuous as they rose and slowly fluttered over the stream, they were seen and snapped up, every one of them, by those fateful sooty-coloured demons of the air, ever streaming by on their swift scythe-shaped wings. Not a swallow nor a martin was in sight at that spot.

It is plain, then, that if the mayfly is declining and dying out because some too greedy bird snatches its life before it can lay its eggs to continue the species, or drop upon the water to supply the trout with its proper succulent food, the swift and not the swallow is the chief culprit.

It is equally plain that these (from the angler's point of view) injurious birds are not breeders by the water-side. Their numbers are too great: they come, ninety per cent. of them I should say, from farm-houses, villages, and towns at a distance of a good many miles from the water.

The revels of the swifts were brought prematurely to an end by a great change in the weather, which began with a thunderstorm on July 27, and two days later a greater storm, with hail the size of big marrowfat peas, which fell so abundantly that the little lawn was all white as if snow had fallen. From that time onwards storm succeeded storm, and finally the weather became steadily bad; and we had rough, cold, wet days right on to the 10th of August. It was a terrible time for the poor holiday people all over the country, and bad too for the moulting and late-breeding birds. As a small set-off to all the discomfort of these dreary days, we had a green lawn once more at the cottage. I had made one or two attempts at watering it, but the labour proved too great to a lazy man, and now Nature had come with her great watering-pot and restored its spring-like verdure and softness.

: During the wettest and coldest days I spent hours watching the swallows and swifts flying about all day long in the rain. These are, indeed, our only summer land birds that never seek a shelter from the wet, and which are not affected in their flight by a wetted plumage. Their upper feathers are probably harder and more closely knit and impervious to moisture than those of other kinds. It may be seen that a swallow or swift, when flying about in the rain, at short intervals gives himself a quick shake as if to throw off the raindrops. Then, too, the food and constant exercise probably serve to keep them warmer than they would be sitting motionless in a dry place. Swifts, we sometimes see, are numbed, and even perish of cold during frosty nights in spring; I doubt that the cold ever kills them by day when they can keep perpetually moving.

Day by day, during this long spell of summer wet and cold, these birds diminished in number, until they were almost all gone—swifts, swallows, and house-martins; but we were not to be without a swallow, for as these went, sand-martins came in, and increased in numbers until they were in thousands. We had them every day and all day before us, flying up and down the valley, in the shelter of the woods, their pale plumage and wavering flight making them look in the distance like great white flies against the wall of black-green trees and gloomy sky beyond.

On days when the sun shone they came in numbers to perch on the telegraph wires stretched across a field between the cottage and village. It was beautiful to see them, a double line fifty or sixty yards long of the small, pale-coloured, graceful birdlings, sitting so close together as to be almost touching, all with their beaks pointing to the west, from where the wind blew.

In this same field, one day when this pleasant company were leaving us after a week's rest, I picked up one that had killed himself by striking against the wire. A most delicate little dead swallow, looking in his pale colouring and softness as moth-like in death as he had seemed when alive and flying. I took him home—the little moth-bird pilgrim to Africa, who had got no farther than the Itchen on his journey—and buried him at the roots of a honeysuckle growing by the cottage door. It seemed fittest that he should be put there, to become part with the plant which, in the pallid yellows and dusky reds of its blossoms, and in the perfume it gives out so abundantly at eventide, has an expression of melancholy, and is more to us in some of our moods than any other flower.

The bad weather brought to our little plot of ground a young blackbird, who had evidently been thrown upon the world too early in life. A good number of blackbird broods had been brought off in the bushes about us, and in the rough and tumble of those tempestuous days some of the young had no doubt got scattered and lost; this at all events was one that had called and called to be fed and warmed and comforted in vain—we had heard him calling for days—and who had now grown prematurely silent, and had soberly set himself to find his own living as best he could. Between the lawn and the small sweetbriar hedge there was a strip of loose mould where roses had been planted, and here the bird had discovered that by turning over the dead leaves and loose earth a few small morsels were to be found. During those cold, windy, wet days we observed him there diligently searching in his poor, slow, little way. He would strike his beak into the loose ground, making a little hop forward at the same time to give force to the stroke, and throw up about as much earth as would cover a shilling-piece; then he would gaze attentively at the spot, and after a couple of seconds hop and strike again; and finally, if he could see nothing to eat, he would move on a few inches and begin again in another place. That was all his art —his one poor little way of getting a living; and it was plain to see from his bedraggled appearance and feeble motions, that he was going the way of most young orphaned birds.

Now, I hate playing at providence among the creatures, but we cannot be rid of pity; and there are exceptional cases in which one feels justified in putting out a helping hand. Nature herself is not always careless of the individual life: or perhaps it would be better to say with Thoreau—"We are not wholly involved in Nature." And anxious to give the poor bird a chance by putting him in a sheltered place, and feeding him up, as Ruskin once did in a like case, I set about catching him, but could not lay hands on him, for he was still able to fly a little, and always managed to escape

AN ORPHANED BLACKBIRD

pursuit among the brambles, or else in the sedges by the waterside. Half-an-hour after being hunted, he would be back on the edge of the lawn prodding the ground in the old feeble, futile way. And the scraps of food I cunningly placed for him he disregarded, not knowing in his ignorance what was good for him. Then I got a supply of small earthworms, and, stalking him, tossed them so as to cause them to fall near him, and he saw and knew what they were, and swallowed them hungrily; and he saw, too, that they were thrown to him by a hand, and that the hand was part of that same huge grey-clad monster that had a little while back so furiously hunted him; and at once he seemed to understand the meaning of it all, and instead of flying from he ran to meet us, and, recovering his voice, called to be fed. The experience of one day made him a tame bird; on the second day he knew that bread and milk, stewed plums, pie-crust, and, in fact, anything we had to give, was good for him; and in the course of the next two or three days he acquired a useful knowledge of our habits. Thus, at half-past three in the morning he would begin calling to be fed at the bedroom window. If no notice was taken of him he would go away to try and find something for himself, and return at five o'clock when breakfast was in preparation, and place himself before the kitchen door. Usually he got a small snack then; and at the breakfast hour (six o'clock) he would turn up at the dining-room window and get a substantial meal. Dinner and tea time—twelve and half-past three o'clock—found him at the same spot; but he was often hungry between meals, and he would then sit before one door or window and call, then move to the next door, and so on until he had been all round the cottage. It was most amusing to see him when, on our return from a long walk or a day out, he would come to meet us, screaming

excitedly, bounding over the lawn with long hops, looking like a miniature very dark-coloured kangaroo.

One day I came back alone to the cottage, and sat
down on the lawn in a canvas chair, to wait for my
companion who had the key. The blackbird had seen,
and came flying to me, and pitching close to my feet
began crying to be fed, shaking his wings, and dancing
about in a most excited state, for he had been left a
good many hours without food, and was very hungry.
As I moved not in my chair he presently ran round and
began screaming and fluttering on the other side of it,
thinking, I suppose, that he had gone to the wrong
place, and that by addressing himself to the back of my
head he would quickly get an answer.

The action of this bird in coming to be fed naturally
attracted a good deal of attention among the feathered
people about us; they would look on at a distance,
evidently astonished and much puzzled at our bird's
boldness in coming to our feet.   But nothing dreadful
happened to him, and little by little they began to lose
their suspicion; and first a robin—the robin is always
first—then other blackbirds to the number of seven,
then chaffinches and dunnocks, all began to grow tame
and to attend regularly at meal-time to have a share in
anything that was going. The most lively, active, and
quarrelsome member of this company was our now
glossy foundling; and it troubled us to think that in
feeding him we were but staving off the evil day when
he would once more have to find for himself. Certainly

we were teaching him nothing. But our fears were idle. The seven wild blackbirds that had formed a habit of coming to share his food were all young birds, and as time went on and the hedge fruit began to ripen, we noticed that they kept more and more together. Whenever one was observed to fly straight away to some distance, in a few moments another would follow, then another; and presently it would be seen that they were all making their way to some spot in the valley, or to the woods on the other side. After several hours' absence they would all reappear on the lawn, or near it, at the same time, showing that they had been together throughout the day and had returned in company. After observing them in their comings and goings for several weeks, I felt convinced that this species has in it the remains of a gregarious instinct which affects the young birds.

Our bird, as a member of this little company, must have quickly picked up from the others all that it was necessary for him to know, and at the last it was plain to us from his behaviour at the cottage that he was doing very well for himself. He was often absent most of the day with the others, and on his return late in the afternoon he would pick over the good things placed for him in a leisurely way, selecting a morsel here and there, and eating more out of compliment to us, as it seemed, than because he was hungry. But up to the very last, when he had grown as hardy and strong on the wing as any of his wild companions, he kept up his acquaintance with and confidence in us; and even at night when I would go out to where most of our wild birds roosted, in the trees and bushes growing in a vast old chalk-pit close to the cottage, and called "Blackie," instantly there would be a response—a softly chuckled note, like a sleepy "Good-night," thrown back to me out of the darkness.

During the spell of rough weather which brought us the blackbird, my interest was centred in the cirl buntings. On August 4, I was surprised to find that they were breeding again in the little sweetbriar hedge, and had three fledglings about a week old in the nest. They had on this occasion gone from the west to the east side of the cottage, and the new nest, two to three feet from the ground, was placed in the centre of a small tangle of sweetbriar, bramble, and bryony, within a few yards of the trunk of the big lime tree under which I was accustomed to sit. I had this nest under observation until August 9, which happened to be the worst day, the coldest, wettest, and windiest of all that wintry spell; and yet in such weather the young birds came out of their cradle. For a couple of days they remained near the nest concealed among some low bushes; then the whole family moved away to a hedge at some distance on higher ground, and there I watched the old birds for some days feeding their young on grasshoppers.

The result of my observations on these birds and on three other pairs which I found breeding close by—one in the village, another just outside of it, and the third by the thorn-grown foundation of ruined Abbotstone not far off—came as a surprise to me; for it appeared that the cirl in its breeding habits and language was not like other buntings, nor indeed like any other bird. The young hatched out of the curiously marked or "written" eggs are like those of the yellow-hammer, black as moor-hen chicks in their black down, opening wide crimson mouths to be fed. But should the parent birds, or one of them, be watching you at the nest, they will open not their beaks, but hearing and obeying the warning note they lie close as if glued to the bottom of the nest. It is a curious sound. Unless one knows it, and the cause of it, one may listen a long time and not discover the bird that utters it. The buntings sit as usual, motionless and unseen among the leaves of the tree, and so long as you are near the nest, keep up the sound, an excessively sharp metallic chirp, uttered in turns by both birds, but always a short note in the female, and a double note in the male, the second one prolonged to a wail or squeal. No other bird has an alarm or warning note like it: it is one of those very high sounds that are easily missed by the hearing, like the robin's fine-drawn wail when in trouble about his young; but when you catch and listen to it the effect on the brain is somewhat distressing. A Hampshire friend and naturalist told me that a pair of these birds that bred in his garden almost drove him crazy with their incessant sharp alarm note.

The effect of this warning sound on the young is very striking: before they can fly or are fit to leave the nest, they are ready, when approached too closely, to leap like startled frogs out of the nest, and scuttle away into hiding on the ground. Once they have flown they are extremely difficult to find, as, on hearing the parent's warning note, they squat down on their perch and remain motionless as a leaf among the leaves. Often I could only succeed in making them fly by seizing and shaking the branches of a thorn or other bush in which I knew they were hidden. So long as the young bird keeps still on its branch, the old bird on some tree twenty or thirty or forty yards away remains motionless, though all the time emitting the sharp, puzzling, warning sound; but the very instant that the young bird quits his perch, darting suddenly away, the parent bird is up too, shooting out so swiftly as almost to elude the sight, and in a moment overtakes and flies with the young bird, hugging it so closely that the two look almost like one. Together they dart away to a distance, usually out over a field, and drop and vanish in the grass. But in a few moments the parent bird is back again, sitting still among the leaves, emitting the shrill sound, ready to dart away with the next young bird that seeks to escape by flight.

This method of attending and safe-guarding the young is, indeed, common among birds, but in no species known to me is it seen in such vigour and perfection. What most strikes one is the change from immobility when the bird sits invisible among the leaves, marking the time with those excessively sharp, metallic clicks and wails like a machine-bird, to unexpected, sudden, brilliant activity.

When not warned into silence and immobility by the parent the young cirls are clamorous enough, crying to be fed, and these, too, have voices of an excessive sharpness. Of other native species the sharpest hunger-cries that I know are those of the tits, especially the long-tailed tit, and the spotted fly-catcher; but these sounds are not comparable in brain-piercing acuteness to those of the young cirls.

Another thing I have wondered at in a creature of so quiet a disposition as the cirl bunting is the extraordinary violence of the male towards other small birds when by chance they come near his young, in or out of the nest. So jealous is he that he will attack a willow-wren or a dunnock with as much fury as other birds use only towards the most deadly enemies of their young.

Here, by the Itchen, where we have all four buntings, I find that the reed bunting—called black-head or black-top—is, after the cirl, the latest singer. He continues when, towards the end of August, the corn bunting and yellow-hammer become silent. He is the poorest singer of the bunting tribe, the first part of his song being like the chirp of an excited sparrow, somewhat shriller, and then follows the long note, shrill too, or sibilant and tremulous. It is more like the distressful hunger-call of some young birds than a song-note. A reedy sound in a reedy place, and one likes to hear it in the green valley among the wind-rustled, sword-shaped leaves and

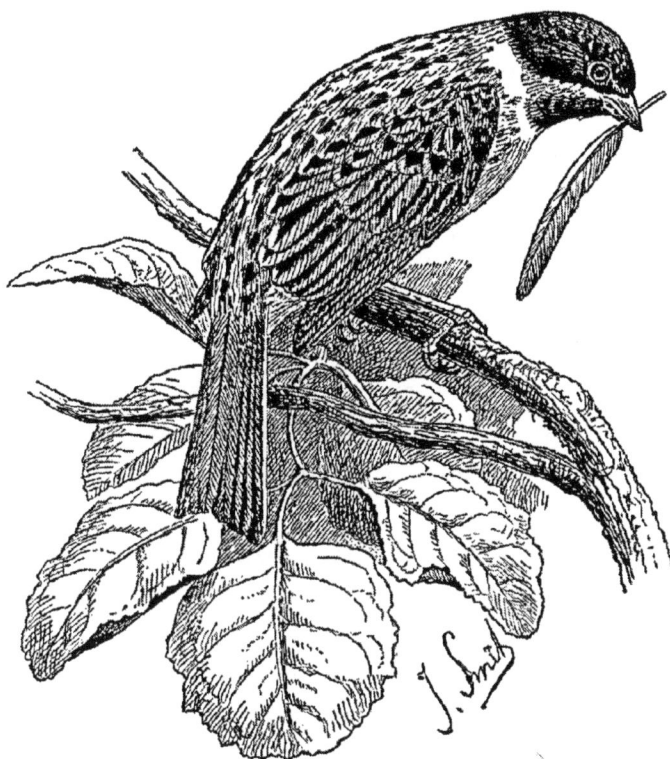

REED BUNTING

waving spears of rush and aquatic grass. So fond is he of his own music that he will sing even when moulting. I was amused one day when listening to a reed bunting sitting on a top branch of a dwarf alder tree in the valley by Ovington, busily occupied in preening his

U

fluffed-out and rather ragged-looking plumage, yet
pausing at short intervals in his task to emit his song.
So taken up was he with the feather-cleaning and
singing, that he took no notice of me when I walked
to within twenty-five yards of him. By-and-by, in
passing one of his long flight-feathers through his
beak it came out, at which he appeared very much surprised. First he raised his head, then began turning
it about this way and that, as if admiring the feather
he held, or trying to get a better sight of it. For
quite a minute he kept it, forgetting to sing, then in
turning it about he accidentally dropped it. Bending
his head down, he watched its slow fall to the grass
below very intently, and continued gazing down even
after it was on the ground; then, pulling himself
together, he resumed the feather-preening task, with
its musical interludes.

The worst day during the bad weather when the
young cirl buntings left the nest brought the wintry
spell to an end. A few days of such perfect weather
followed that one could wish for no higher good than
to be alive on that green earth, beneath that blue sky.
One could best appreciate the crystal purity and divine
blueness of the immense space by watching the rooks
revelling on high in the morning sunshine, looking in
their blackness against the crystalline blue like bird-
figures with outspread, motionless wings, carved out of
anthracite coal, and suspended by invisible wires in
heaven. You could watch them, a numerous company,

moving upward in wide circles, the sound of their voices
coming fainter and fainter back to earth, until at that
vast height thêy seemed no bigger than humble-bees

This clarity of the atmosphere had a striking effect,
too, on the appearance of the trees, and I could not
help noticing the superiority of the oak to all other
forest trees in this connection.   There comes a time in
late summer when at last it loses that "glad light
grene" which has distinguished it among its dark-
leafed neighbours, and made it in our eyes a type of
unfading spring and of everlastingness.   It grows dark,
too, at last, and is as dark as a cypress or a cedar of
Lebanon; but observe how different this depth of
colour is from that of the elm.   The elm, too, stands
alone, or in rows, or in isolated groups in the fields, and
in the clear sunshine its foliage has a dull, summer-
worn, almost rusty green.   There is no such worn and
weary look in the foliage of the oak in August and
September.   It is of a rich, healthy green, deep but
undimmed by time and weather, and the leaf has a
gloss to it.   Again, on account of its manner of growth,
with widespread branches and boughs and twigs well
apart, the foliage does not come before us as a mere
dense mass of green—an intercepting cloud, as in a
painted tree; but the sky is seen through it, and
against the sky are seen the thousand thousand individual leaves, clear cut and beautiful in shape.

It was one of my daily pleasures during this fine
weather to go out and look at one of the solitary oak

trees growing in the adjoining field when the morning sunlight was on it. To my mind it looked best when viewed at a distance of sixty to seventy yards across the open grass field with nothing but the sky beyond. At that distance not only could the leaves be distinctly seen, but the acorns as well, abundantly and evenly distributed over the whole tree, appearing as small globes of purest bright apple-green among the deep green foliage. The effect was very rich, as of tapestry with an oak-leaf pattern and colour, sprinkled thickly over with round polished gems of a light-green sewn into the fabric.

To an artist with a soul in him, the very sight of such a tree in such conditions would, I imagined, make him sick of his poor little ineffectual art.

CHAPTER XIV

THE oak in the field and a flowering plant by the water were the two best things plant life contained for me during those beautiful late summer days by the Itchen. About the waterside flower I must write at some length.

Of our wild flowers the yellow in colour, as a rule, attract me least; not because the colour is not beautiful to me, but probably on account of the numerous ungraceful, weedy-looking plants of unpleasant scent which in late summer produce yellow flowers—tansy, fleabane, ragwort, sow-thistle, and some of other orders, the worst of the lot being the pepper saxifrage, an ungainly parsley in appearance, with evil-smelling flowers. You know them by their odours. If I were to smell at a number of strong-scented flowers unknown to me in a dark room, or blindfolded, I should be able to pick out the yellow ones. They would have the yellow smell. The yellow smell has an analogue in the purple taste. It may be fancy, but it strikes me that there is a certain family resemblance in the flavours of most purple fruits, or their skins — the purple fruit-flavour which is so strong in damson, sloe, black currant, blackberry, mulberry, whortleberry, and elderberry.

All the species I have named were common in the valley, and there were others—St. John's-wort, yellow loosestrife, &c. — which, although not ungraceful nor evil-smelling, yet failed to attract. Nevertheless, as the days and weeks went on, and brought yet another conspicuous yellow waterside flower into bloom, which became more and more abundant as the season advanced, while the others, one by one, faded and failed from the earth, until, during the last half of September, it was in its fullest splendour, I was completely won by it, and said in my haste that it was the brightest blossom in all the Hampshire garland, if not the loveliest wild flower in England. Nor was it strange, all things considered, that I was so taken with its beauty, since, besides being beautiful, it was new to me, and therefore had the additional charm of novelty; and, finally, it was at its best when all the conspicuous flowers that give touches of brilliant colour here and there to the green of this greenest valley, including most of the yellow flowers I have mentioned, were faded and gone.

No description of this flower, Mimulus luteus, known

to the country people as "wild musk," is needed here
—it is well known as a garden plant. The large
foxglove-shaped flowers grow singly on their stems
among the topmost leaves, and the form of stem,
leaf, and flower is a very perfect example of that

MIMULUS LUTEUS

kind of formal beauty in plants which is called
"decorative." This character is well shown in the above
figure, reduced to little more than half the natural
size, from a spray plucked at Bransbury, on the Test.
But the shape is nothing, and is scarcely seen or
noticed twenty-five to fifty yards away, the proper

distance at which to view the blossoming plants; not indeed as a plant-student or an admirer of flowers in a garden would view it, as the one thing to see, but merely as part of the scene. The colour is then everything. There is no purer, no more beautiful yellow on any of our wild flowers, from the primrose and the almost equally pale, exquisite blossom which we improperly name "dark mullein" in our books on account of its lovely purple eye, to the intensest pure yellow of the marsh marigold.

But although purity of colour is the chief thing, it would not of itself serve to give so great a distinction to this plant; the charm is in the colour and the way in which Nature has disposed it, abundantly, in single, separate blossoms, among leaves of a green that is rich and beautiful, and looks almost dark by contrast with that shining, luminous hue it sets off so well.

On September 17 it was Harvest Festival Sunday at the little church at Itchen Abbas, where I worshipped that day, and I noticed that the decorators had dressed up the font with water-plants and flowers from the river; reeds and reed-mace, or cat's-tail, and the yellow mimulus. It was a mistake. Deep green, glossy foliage, and white and brilliantly coloured flowers look well in churches; white chrysanthemums, arums, azaleas, and other conspicuous white flowers; and scarlet geraniums, and many other garden blooms which seen in masses in the sunshine hurt the sense

—cinerarias, calceolarias, larkspurs, &c. The subdued
light of the interior softens the intensity, and sometimes crudity, of the strongest colours, and makes
them suitable for decoration. The effect is like that
of stained-glass windows, or of a bright embroidery
on a sober ground. The graceful, grey, flowery reeds,
and the light-green reed-mace, with its brown velvet
head, and the moist yellow of the mimulus, which
quickly loses its freshness, look not well in the dim,
religious light of the old village church. These should
be seen where the sunlight and wind and water are,
or not seen at all.

Beautiful as the mimulus is when viewed in its
natural surroundings, by running waters amidst the
greys and light and dark greens of reed and willow,
and of sedge and aquatic grasses, and water-cress, and
darkest bulrush, its attractiveness was to me greatly
increased by association. Now to say that a flower
which is new to one can have any associations may
sound very strange, but it is a fact in this case.
Viewing it at a distance of, say, forty or fifty yards, as
a flower of a certain size, which might be any shape,
in colour a very pure, luminous yellow, blooming in
profusion all over the rich green, rounded masses of
the plants, as one may see it in September at Ovington, and at many other points on the Itchen, from
its source to Southampton Water, and on the Test,
I am so strongly reminded of the yellow camaloté
of the South American watercourses that the memory

is almost like an illusion. It has the pure, beautiful
yellow of the river camaloté; in its size it is like
that flower; it grows, too, in the same way, singly,
among rounded masses of leaves of the same lovely
rich green; and the camaloté, too, has for neighbours the green blades of the sedges, and grey, graceful
reeds, and multitudinous bulrushes, their dark polished
stems tufted with brown.

Looking at these masses of blossoming mimulus
at Ovington, I am instantly transported in thought to
some waterside thousands of miles away. The dank,
fresh smell is in my nostrils; I listen delightedly
to the low, silvery, water - like gurgling note of
the little kinglet in his brilliant feathers among the
rushes, and to the tremulous song of the green marsh
grasshoppers or leaf crickets; and with a still greater
delight do I gaze at the lovely yellow flower, the unforgotten camaloté, which is as much to me as the wee,
modest, crimson-tipped daisy was to Robert Burns or
to Chaucer; and as the primrose, the violet, the dog-
rose, the shining, yellow gorse, and the flower o' the
broom, and bramble, and hawthorn, and purple heather
are to so many inhabitants of these islands who were
born and bred amid rural scenes.

On referring to the books for information as to
the history of the mimulus as a British wild flower,
I found that in some it was not mentioned, and in
others mentioned only to be dismissed with the re-
mark that it is an "introduced plant." But when was

it introduced, and what is its range ? And whom' are
we to ask ?

After an infinite amount of pains, seeing and writing to all those among my acquaintance who have
any knowledge of our wild plant life, I discovered
that the mimulus grows more or less abundantly in
or by streams here and there in most English counties,
but is more commonly met with south of Derby-
shire ; also that it extends to Scotland, and is known
even in the Orkneys. Finally, a botanical friend discovered for me that as long ago as 1846 there had
been a great discussion, in which a number of persons
took part, on this very subject of the date of the
naturalisation in Britain of the mimulus, in Edward
Newman's botanical magazine, the *Phytologist*. It was
shown conclusively by a correspondent that the· plant
had established itself at one point as far back as the
year 1815.

There may exist more literature on the subject if one
knew where to look for it ; but we are certainly justified
in feeling annoyed at the silence of the makers of books
on British wild flowers, and the compilers of local lists
and floras. And *what*, we should like to ask of our
masters, is a British wild flower ? Does not the same
rule apply to plants as to animals—namely, that when a
species, whether " introduced " or imported by chance
or by human agency, has thoroughly established itself
on our soil, and proved itself able to maintain its
existence in a state of nature, it becomes, and is a

British species? If this rule had not been followed by zoologists, even our beloved little rabbit would not be a native, to say nothing of our familiar brown rat and our black-beetle: and the pheasant, and red-legged partridge, and capercailzie, and the fallow-deer, and a frog, and a snail, and goodness knows how many other British species, introduced into this country by civilised man, some in recent times. And, going farther back in time, it may be said that every species has at some time been brought, or has brought itself from otherwhere—every animal from the red deer and the white cattle, to the smallest, most elusive microbe not yet discovered; and every plant from the microscopical fungus to the British oak and the yew. The main thing is to have a rule in such a matter, a simple, sensible rule, like that of the zoologist, or some other; and what we should like to know from the botanists is —Have they got a rule, and, if so, what is it? There are many who would be glad of an answer to this question: judging from the sale of books on British wild flowers during the last few years, there must be several millions of persons in this country who take an interest in the subject.

One bright September day, when the mimulus was in its greatest perfection, and my new pleasure in the flower at its highest, I by chance remembered that Gilbert White, of Selborne, in the early part of his career, had been curate for a time at Swarraton, a small village on the Itchen, near its source, about four

miles above Alresford. That was in 1747. To Swarraton
I accordingly went, only to find what any guide-book
or any person would have told me, that the church no
longer exists. Only the old churchyard remained,
overgrown with nettles, the few tombstones that had
not been carried away so covered with ivy as to appear
like green mounds, A group of a dozen yews marked
the spot where the church had formerly stood; and
there were besides some very old trees, an ancient yew
and a giant beech, and others, and just outside the
ground as noble an ash-tree as I have ever seen. These
three, at any rate, must have been big trees a century
and a half ago, and well known to Gilbert White. On
inquiry I was told that the church had been pulled
down a very long time back—about forty years, perhaps; that it was a very old and very pretty church,
covered with ivy, and that no one knew why it was
pulled down. The probable reason was that a vast
church was being or about to be built at the neighbouring village of Northington, big enough to hold all the
inhabitants of the two parishes together, and about a
thousand persons besides. This immense church would
look well enough among the gigantic structures of all
shapes and materials in the architectural wonderland
of South Kensington. But I came not to see this
building: the little ancient village church, in which the
villagers had worshipped for several centuries, where
Gilbert White did duty for a year or so, was what I
wanted, and I was bitterly disappointed. Looking

away from the weed-grown churchyard, I began to
wonder what his feelings would be could he revisit this
old familiar spot. The group of yew-trees where the
church had stood, and the desolate aspect of the ground
about it would disturb and puzzle his mind; but, on
looking farther, all the scene would appear as he had
known it so long ago—the round, wooded hills, the
green valley, the stream, and possibly some of the old
trees, and even the old cottages. Then his eyes would
begin to detect things new and strange. First, my
bicycle, leaning against the trunk of the great ash-tree,
would arrest his attention; but in a few moments,
before he could examine it closely and consider for
what purpose it was intended, something far more
interesting and more wonderful to him would appear in
sight. Five large birds standing quietly on the green
turf beside the stream—birds never hitherto seen.
Regarding them attentively, he would see that they
were geese, and it would appear to him that they were
of two species, one white and grey in colour, with black
legs, the other a rich maroon red, with yellow legs;
also that they were both beautiful and more graceful in
their carriage than any bird of their family known to
him. Before he would cease wondering at the presence
at Swarraton of these Magellanic geese, no longer
strange to any living person's eyes in England, lo! a
fresh wonder—beautiful yellow flowers by the stream,
unlike any flower that grew there in his day, or by any
stream in Hampshire.

But how long after White's time did that flower run
wild in Hampshire? I asked, and then thought that I
might get the answer from some old person who had
spent a long life at that spot.

I went no farther than the nearest cottage to find
the very one I wanted, an ancient dame of seventy-
four, who had never lived anywhere but in that small
thatched cottage at the side of the old churchyard.
She was an excessively thin old dame, and had the
appearance of a walking skeleton in a worn old cotton
gown; and her head was like a skull with a thin grey
skin drawn tightly over the sharp bones of the face,
with pale-coloured living eyes in the sockets. Her
scanty grey hair was gathered in a net worn tightly
on her head like a skull-cap. The old women in
the villages here still keep to this long-vanished
fashion.

I asked this old woman to tell me about the yellow
flowers by the water, and she said that they had always
been there. I told her she must be mistaken; and
after considering for awhile she assured me that they
grew there in abundance when she was quite young.
She distinctly remembered that before her marriage—
and that was over fifty years ago—she often went
down to the stream to gather flowers, and would come
in with great handfuls of wild musk.

When she had told me this, even before she had
finished speaking, I seemed to see two persons before
me — the lean old woman with her thin colourless

visage, and, coming in from the sunshine, a young
woman with rosy face, glossy brown hair and laughing
blue eyes, her hands full of brightest yellow wild musk
from the stream.   And the visionary woman seemed

Wild Musk.

to be alive and real, and the other unsubstantial, a
delusion of the mind, a ghost of a woman.

But was the old woman right—was the beautiful
yellow mimulus, the wild musk or water buttercup
as she called it, which our botanists refuse to admit
into their works intended for our instruction, or give

it only half-a-dozen dry words—was it a common
wild flower on the Hampshire rivers more than half
a century ago?

From the valley and the river with its shining
yellow mimulus and floating water - grass in the
crystal current—that green hair-like grass that one
is never tired of looking at—back to the ivy-green
cottage, its ancient limes and noble solitary oaks, and,
above all, its birds; then back again to the stream—
that mainly was our life. But close by on either side
of the valley were the downs, and these too drew us
with that immemorial fascination which the higher
ground has for all of us, because of the sense of
freedom and power which comes with a wide horizon.
That was a fine saying of Lord Herbert of Cherbury
that a man mounted on a good horse is lifted above
himself: one experiences the feeling in a greater degree
on any chalk down. One extensive open down within
easy distance was a favourite afternoon walk. Here
on the short fragrant turf an army of peewits were to
be found every day, and usually there were a few
stone curlews with them. It is not here as in the
country about Salisbury, where the Hawking Club has
its headquarters, and where they have been "having
fun with the thick-knees," as they express it in their
lingo, until there are no thick-knees left. But the
chief attraction of this down was an extensive thicket
of thorn and bramble, mixed with furze and juniper

x

and some good-sized old trees, where birds were abundant, many of them still breeding. Here, down to the end of September, I found turtle-doves' nests with newly-hatched young and incubated eggs. I always felt more than compensated for scratches and torn clothes when I found young turtle-doves in the down, as the little creatures are then delightful to look at. Sitting hunched up on its platform, the head with its massive bulbous beak drawn against its arched back, the little thing is less like a bird than a mammal in appearance—a singularly coloured shrew, let us say. The colour is indeed strange, the whole body, the thick, fleshy, snout-like beak included, being a deep, intense, almost indigo blue, and the loose hair-like down on the head and upper parts a light, bright primrose yellow.

There are surprising colours in some young birds: the cirl nestling, as we have seen, is black and crimson —clothed in black down with gaping crimson mouth; loveliest of all is the young snipe in down of brown-gold, frosted with silvery white ; but for quaintness and fantastic colouring the turtle-dove nestling has no equal. In all of our native doves, and probably in all doves everywhere, the skin is blue and the down yellow, but the colours differ in intensity. I tried to find a newly-hatched stock-dove to compare it with the turtle nestling but failed, although the species is quite common and, like the other two, breeds till October. Ring-dove nestlings were easy to see, but in these the blue colour, though deep on the beak

and head, is quite pale on the body, fading almost to
white on some parts; and the down, too, is very pale,
fading to whitish tow-colour on the sides and back.

When seeking for a ring-dove in down I had an
amusing adventure. At a distance of some miles from
the Itchen, near the Test, one day in September, I was
hunting for an insect I wanted in a thick copse by
Tidbury Ring, an ancient earthwork on the summit
of a chalk hill. Hearing a boy's voice singing near, I

peeped out and saw a lad of about fifteen tending some
sheep: he was walking about on his knees, trimming
the herbage with an old rusty pair of shears which
he had found! It startled him a little when I burst
out of the cover so near him, but he was ready to
enter into conversation, and we had a long hour together, sitting on the sunny down. I mentioned my
desire to find a newly-hatched ring-dove, and he at
once offered to show me one. There were two nests
with young close by, in one the birds were half-fledged,
the others only came out of their shells two days

before. These we went to look for, the boy leading the way to a point where the trees grew thickest. He climbed a yew, and from the yew passed to a big beech tree, in which the nest was placed, but on getting to it he cried out that the nest was forsaken and the young dead. He threw them down to me, and he was grieved at their death as he had known about the nest from the time it was made, and had seen the young birds alive the day before. No doubt the parents had been shot, and the cold night had quickly killed the little ones.

This was the most intelligent boy I have met in Hampshire; he knew every bird and almost every insect I spoke to him about. He was, too, a mighty hunter of little birds, and had captured stock-doves and wheatears in the rabbit burrows. But his greatest feat was the capture of a kingfisher. He was down by the river with a sparrow-net at a spot where the bushes grow thick and close to the water, when he saw a kingfisher come and alight on a dead twig within three yards of him. The bird had not seen him standing behind the bush : it sat for a few moments on the twig, its eyes fixed on the water, then it dropped swiftly down, and he jumped out and threw the net over it just as it rose up with a minnow in its beak. He took it home and put it in a cage.

I gave him a sharp lecture on the cruelty of caging kingfishers, telling him how senseless it was to confine such a bird, and how impossible to keep it alive in prison. It was better to kill them at once if he wanted to destroy them. "Of course your kingfisher died," I said.

"No," he replied. He stood the cage on a chair, and the bird was no sooner in it than his little sister, a child of two who was fidgeting round, pulled the door open and out flew the kingfisher!

Returning to the cottage, whether from the high down, the green valley, or the silent, shady wood, it always seemed a favourite dwelling or nesting-place of the birds, where indeed they most abounded. Now that bright genial weather had come after the cold and storm to make them happy, the air was full of their chirpings and twitterings, their various little sounds of conversation and soliloquy, with an occasional bright, loud, perfect song. It was generally the wren, whose lyric changes not through all the changeful year, that uttered it. It was this small brown bird, too, that amused me most with the spectacle of his irrepressible delight in the new warmth and sunlight. There were about a dozen wrens at the cottage, and some of them were in the habit of using their old undamaged nests in the ivy and woodbine as snug little dormitories. But they cared nothing for the human inhabitants of the cottage; they were like small birds that had built their nest in the interstices of an eagles' eyrie, who knew nothing and cared nothing about the eagles. Occasionally, when a wren peeped in from the clustering ivy or hopped on to a window-sill and saw us inside, he would scold us for being there with that sharp, angry little note of his, and then fly away. Nor would he take a crumb from the table spread out of doors every day for the birds that disdained not to be fed. The ivy and

A WREN'S SUN-BATH

creepers that covered the cottage abounded with small spiders, caterpillars, earwigs, chrysalids, and what not; that was good enough for him—Thank you for your kind intentions!

Looking from a window at a bed of roses a few feet away, I discovered that the wren took as much pleasure in a dust bath as any bird. He would come to the loose soil and select a spot where the bed sloped towards the sun, and then wriggle about in the earth with immense enjoyment. Dusting himself, he would look like a miniature partridge with a round body not much bigger than a walnut. After dusting would come the luxurious sun-bath, when, with feathers raised and minute wings spread out and beak gaping, the little thing would lie motionless and panting; but at intervals of three or four seconds a joyful fit of shivering would seize him, and at last, the heat becoming too great, he would shake himself and skip away, looking like a brown young field vole scuttling into cover.

This bright and beautiful period came to an end on August 22, and we then had unsettled weather with many sudden changes until September 3—cloudy oppressive days, violent winds, thunderstorms, and days of rain and sunshine, and morning and evening rainbows; it was a mixture of April, midsummer, and October.

This changeful period over, there was fine settled weather; it was the golden time of the year, and it continued till our departure on the last day of September.

The fruit season was late this year—nearly a fortnight later than in most years; and when the earliest, the wild arum, began to ripen, the birds—thrushes and chaffinches were detected—fell upon and devoured all the berries, regardless of their poisonous character, almost before their light-green had changed to vivid scarlet. Then came the deep crimson fruit of the

honeysuckle; it ripened plentifully on the plants growing against the cottage, and the cole-tits came in bands
to feed on it. It was pretty to see these airy little acro-
bats clinging to the twine-like pendent sprays hanging
before an open window or door. They were like the
little birds in a Japanese picture which one has
seen. Then came the elderberries, which all fruit-
loving birds feast on together. But the tits and finches
and warblers and thrushes were altogether out-num-
bered by the starlings that came in numbers from the
pasture-lands to take part in the great fruit-feast.

The elder is a common tree here, but at the cottage
we had, I think, the biggest crop of fruit in the neighbourhood; and it now occurs to me that the vast old
chalk pit in which the trees grew has not yet been
described, and so far has only been once mentioned
incidentally. Yet it was a great place, but a few yards
away at the side of the old lime trees and the small protecting fence. The entrance to it and its wide floor was
on a level with the green valley, while at its upper end
it formed a steep bank forty feet high. It was doubtless a very old pit, with sides which had the appearance
of natural cliffs and were overhung and draped with
thorn-trees, masses of old ivy, and traveller's joy. Inside it was a pretty tangled wilderness; on the floor
many tall annuals flourished—knapweed and thistle
and dark mullein and teazel, six to eight feet high.
Then came some good-sized trees—ash and oak, and
thorn, bramble and elder in masses. It was a favourite

breeding-place of birds of many species; even the red-backed shrike had nested there within forty yards of a human habitation, and the kingfisher had safely reared his young, unsuspected by the barbarous water-keeper. The pit, too, was a shelter in cold rough weather and a roosting-place at night. Now the fruit was ripe, it was a banqueting-place as well, and the native birds were joined by roving outsiders, missel-thrushes in scores, and starlings in hundreds. The noise they produced— a tangle of so many various semi-musical voices— sounded all day long; and until the abundant fruit had all been devoured the chalk pit was a gigantic green and white bowl full to overflowing with sunshine, purple juice, and melody.

The biggest crop of this fruit out of the old chalk pit was in the garden of a cottage in the village, close to the river, occupied by an old married couple, hard workers still with spade and hoe, and able to make a living by selling the produce of their garden. It was a curious place; fruit trees and bushes, herbs, vegetables, flowers, all growing mixed up anyhow, without beds or walks or any line of demarcation between cultivated plants and brambles and nettles on either side and the flags and sedges at the lower end by the river. In the midst of the plot, just visible among the greenery, stood the small, old, low-roofed thatched cottage, where the hens were free to go in and lay their eggs under the bed, or in any dark corner they preferred. A group of seven or eight old elder-trees grew close to the cottage, their branches bent and hanging with the weight of the purpling clusters.

"What are you going to do with the fruit?" I asked the old woman, and this innocent question raised a tempest in her breast, for I had unwittingly touched on a sore subject.

"Do!" she exclaimed rather fiercely, " I'm going to do nothing with it! I've made elderberry wine years and years and years. So did my mother; so did my grandmother; so did everybody in *my* time. And very good it were, too, I tell 'e, in cold weather in winter, made hot. It warmed your inside. But nobody wants it now, and nobody'll help me with it. How 'm I to do it—keep the birds off and all! I've been fighting 'em years and years, and now I can't do it no longer. And what's the good of doing it if the wine's not good enough for people to drink? Nothing's good enough now unless you buys it in a public-house or a shop. It wasn't so when I were a girl. We did everything for ourselves then, and it were better, I tell 'e. We kep' a pig then—so did every one; and the pork and bacon it were good, not like what we buy now. We put it mostly in brine, and let it be for months; and when we took it out and biled it, it were red as a cherry and white as milk, and it melted just like butter in your mouth. That's what we ate in *my* time. But you can't keep a pig now—oh dear, no! You don't have him more 'n a day or two before the sanitary man looks in. He says he were passing and felt a sort of smell about—would you mind letting him come in just to have a sniff round? He expects it might be a pig you've got. In my time we didn't think a pig's smell hurt nobody. They've got their own smell, pigs have, same as dogs and everything else. But we've got very partickler about smells now.

  "And we didn't drink no tea then. Eight shillings a pound, or may be seven-and-six—dear, dear, how was we to buy it! We had beer for breakfast and it did us good. It were better than all these nasty cocoa stuffs we drink now. We didn't buy it at the public-house—we brewed it ourselves. And we had a brick oven then, and could put a pie in, and a loaf, and whatever we wanted, and it were proper vittals. We baked barley bread, and black bread, and all sorts of bread, and it did us good and made us strong. These iron ranges and stoves we have now—what's the good o' they? You can't bake bread in 'em. And the wheat bread you gits from the shop, what's it good for? 'Tisn't proper vittals—it fills 'e with wind. No, I say, I'm not going to git the fruit—let the birds have it! Just look at the greedy things—them starlings! I've shouted, and thrown sticks and all sorts of things, and shaken a cloth at 'em, and it's like calling the fowls to feed. The more noise I make the more they come. What I say is, If I can't have the fruit I wish the blackbirds 'ud git it. People say to me, 'Oh, don't talk to me about they blackbirds—they be the worst of all for fruit.' But I never minded that—because—well, I'll tell 'e. I mind when I were a little thing at Old Alresford, where I were born, I used to be up at four in the morning, in summer, listening to the blackbirds. And mother she used to say, 'Lord, how she do love to hear a blackbird!' It's always been the same. I's always up at four, and in summer I goes out to hear the blackbird when it do sing so beautiful. But them starlings that come messing about, pulling the straws out of the thatch, I've no patience with they. We didn't have so many starlings when I were young. But things is very different now ; and what I say is, I wish they wasn't—I wish they was the same as when I were a girl. And I wish I was a girl again."

Listening to this tirade on the degeneracy of modern times, it amused me to recall the very different feeling on the same subject expressed by the old Wolmer Forest woman. But the Itchen woman had more vigour, more staying-power in her: one could see it in the fresh colour in her round face, and the pure colour and brightness of her eyes—brighter and bluer than in most blue-eyed girls. Altogether, she was one of the best examples of the hard-headed, indomitable Saxon peasants I have met with in the south of England. She was past seventy, impeded by an old infirmity, the mother of many men and women with big families of their own, all scattered far and wide over the county, —all too poor themselves to help her in her old age, or to leave their work and come such a distance to see her, excepting when they were in difficulties, for then they would come for what she could spare them out of her hardly-earned little hoard.

I admired her "fierce volubility"; but that sudden softening at the end about the blackbird's beautiful voice, and that memory of her distant childhood, and her wish, strange in these weary days, to have her hard life to live over again, came as a surprise to me.

In days like these, so bright and peaceful, one thinks with a feeling of wonder that many of our familiar birds are daily and nightly slipping away, decreasing gradually in numbers, so that we scarcely miss them. By the middle of September the fly-catchers and several of the warblers, all but a few laggards, have left us. Even the swallows begin to leave us before that date. On the 8th many birds were congregated at a point on the river a little above the village, and on the 10th a considerable migration took place. Near the end of a fine day a big cloud came up from the north-west, and beneath it, at a good height, the birds were seen flying down the valley in a westerly direction. I went, out and watched them for half-an-hour, standing on the little wooden bridge that spans the stream. They went by in flocks of about eighty to a couple of hundred birds, flock succeeding flock at intervals of three or four minutes. By the time the sun set the entire sky was covered by the black cloud, and there was a thick gloom on the earth; it was then some eight or ten minutes after the last flock, flying high, had passed twittering on its way that a rush of birds came by, flying low, about on a level with my head as I stood leaning on the handrail of the bridge. I strained my eyes in vain to make out what they were—swallows or martins—as in rapid succession, and in twos or threes, they came before me, seen vaguely as dim spots, and no sooner seen than gone, shooting past my head with amazing velocity and a rushing sound, fanning my face with the wind they created, and some of them touching me with their wing-tips.

On the evening of September 18 a second migration was witnessed at the same spot, flock succeeding flock until it was nearly dark. On the following evening, at another point on the river at Ovington, I witnessed a third and more impressive spectacle. The valley spreads out there to a great width, and has extensive beds of reeds, bulrushes, and other water plants, with clumps and rows of alders and willows. It was growing dark; bats were flitting round me in numbers, and the trees along the edge of the valley looked black against the pale amber sky in the west, when very suddenly the air overhead became filled with a shrill confused noise, and, looking up through my binocular, I saw at a considerable height an immense body of swallows travelling in a south-westerly direction. A very few moments after catching sight of them they paused in their flight, and, after remaining a short time at one point, looking like a great swarm of bees, they began rushing wildly about, still keeping up their shrill excited twittering, and coming lower and lower by degrees; and finally, in batches of two or three hundred birds, they rushed down like lightning into the dark reeds, shower following shower of swallows at intervals of two or three seconds, until the last had vanished and the night was silent again.

It was time for them to go, for though the days were warm and food abundant, the nights were growing cold.

The early hours are silent, except for the brown owls that hoot round the cottage from about four o'clock until dawn. Then they grow silent, and the morning is come, cold and misty, and all the land is hidden by a creeping white river mist. The sun rises, and is not seen for half-an-hour, then appears pale and dim, but grows brighter and warmer by degrees; and in a little while, lo! the mist has vanished, except for a white rag, clinging like torn lace here and there to the valley reeds and rushes. Again, the green earth, wetted with mist and dews, and the sky of that soft pure azure of yesterday and of many previous days. Again the birds are vocal; the rooks rise from the woods, an innumerable cawing multitude, their voices filling the heavens with noise, as they travel slowly away to their feeding-grounds on the green open downs; the starlings flock to the bushes, and the feasting and chatter and song begin that will last until evening. The sun sets crimson, and the robins sing in the night and silence. But it is not silent long; before dark the brown owls begin hooting, first in the woods, then fly across to the trees that grow beside the cottage, so that we may the better enjoy their music. At intervals, too, we hear the windy sibilant screech of the white owl across the valley. Then the wild cry of the stone-curlew is heard as the lonely bird wings his way past, and after that late voice there is perfect silence, with starlight or moonlight.

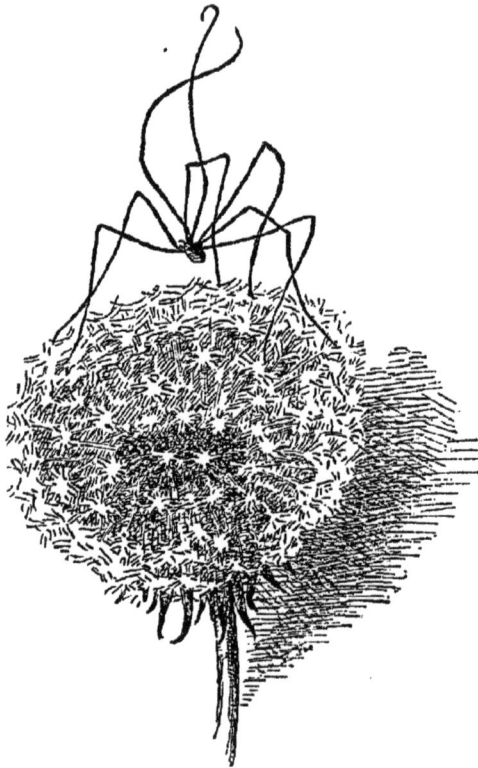

INDEX

Printed by BALLANTYNE, HANSON & Co.
Edinburgh & London

Milton Keynes UK
Ingram Content Group UK Ltd.
UKHW022103150124
436101UK00005B/148